IDEA to PRODUCT:
DIY INVENTION COMMERCIALIZATION

GREAT IDEAS! YOU JUST NEVER KNOW WHEN THEY WILL COME TO YOU!

Do you have an idea for a great new product, or even a garage full of prototypes? Inventing is fun at first, but the overwhelming business aspects can deter you from ever getting the product into the hands of the consumer.

Learn from Dr. Mesut Varlioglu's firsthand experience of converting an idea into a marketable product. Using real-life examples, Dr. Varlioglu explains the best ways for you to patent, manufacture, market, and sell your product to realize the most success in today's marketplace.

Mesut Varlioglu, PhD, is a technical consultant supporting new technology operations across many countries. Through his unique experience, he has learned the tricks of the manufacturing trade. With this practical and comprehensive guide, you can get your idea into the hands of the customers.

IDEA to PRODUCT: DIY INVENTION COMMERCIALIZATION

MESUT VARLIOGLU PhD

MESUT VARLIOGLU PhD

IDEA TO PRODUCT

DIY INVENTION COMMERCIALIZATION

Mesut Varlioglu PhD

After watching the famous TV show *Shark Tank*, you came up with a new product idea. So what? Anyone can come up with something new. Maybe you have discussed your idea with your friends, and they like it. This is encouraging, but anyone can create something that is likable. The important question is whether the idea can be manufactured, marketed, and sold as a product in the marketplace. Most importantly, you need to make a profit by selling the product. This book is about converting an idea to an invention and getting it to the marketplace in the most efficient and profitable way. It explains the process of transforming an idea into a commercial product: how to develop an invention, assess the market, create a prototype, choose a low-cost manufacturing process, and determine price and profit. It also addresses ways you can sell your product and attract distributors. This book is written for engineers, scientists, project managers, entrepreneurs, inventors, and anyone who wants to know the whole process of turning an idea into a commercial product.

ISBN: 1535152761
ISBN 13: 9781535152761
Library of Congress Control Number: 2016911066
CreateSpace Independent Publishing Platform
North Charleston, South Carolina

CONTENTS

ABOUT THE AUTHOR

D r. Mesut Varlioglu is a technical consultant with experience in the defense, aerospace, automotive, and electronics industries. He received his PhD in materials science from Iowa State University in 2009, his MS in materials engineering from Illinois Institute of Technology in 2005, and his BS in metallurgical engineering from Istanbul Technical University of Turkey in 2000. He has authored a number of patents, trademarks, and technical articles in peer-reviewed journals, including *Light Metal Age* and *Journal of Applied Physics*. Additionally, Dr. Varlioglu has given technical seminars and training in the United States, India, China, Taiwan, Germany, and Canada. Most recently, he has commercialized his invention, Measudrill.

INTRODUCTION

I think there is a world market for maybe five computers.
—THOMAS WATSON, PRESIDENT OF IBM, 1943

The world didn't need many computers in 1943. The first computers were the size of houses, and only scientists could use them to sort numbers and complete simple calculations. The world has changed since then, along with the size of computers and their many applications. Today, there are more than one billion computers worldwide (Meulen, 2008), and the calculator is now a simple function on our smartphones.

Since 1943, the world has seen many technological breakthroughs and their applications in new products. Many of these new products were produced locally and focused on the local market. With the collapse of the Soviet Union in the 1990s, a global market emerged, with a massive number of opportunities and customers. To address the needs of these new customers, computers provided a faster communication platform to connect countries and cultures and exchange ideas and products. As computers became more affordable, more people started using them and later started to exchange information via a system called the Internet. I remember getting my first e-mail address in 1996. How exciting it was to read the newspaper without having to buy it or to check information on the Internet or to write e-mails to friends.

We now live in an information age. We are more globally connected to one another than ever before. Information travels as fast as your Internet connection speed, and we are introduced to new products and services almost daily. Currently, almost 250,000 new products are introduced annually to the global market (McInnes, 2014).

Owing to globalization, the consumer marketplace is a price-competitive and volatile environment. As a result, manufacturing, supply chain, and outsourcing operations have moved to countries with more cost-effective labor. A company in the United States now has to compete with manufacturers all over the world in terms of cost. Take a look around your house: Where were the items in your everyday life manufactured? From the computers we use to the toys that our kids play with, they are most likely all from China. For consumer markets with massive production volumes, outsourcing labor has reduced the production cost, thereby creating cheaper products in spite of shipping. Asian manufacturers use networks like Alibaba (www.Alibaba.com) to quickly serve customers and distributors. These organized networks for marketing goods have further increased competition among Asian and US manufacturers.

In recent years, rising labor costs in China have transformed the global market into a fast-changing and extremely volatile and price-competitive manufacturing operation (Maslow, 2015). Companies are now looking for alternative low-cost countries in which to manufacture goods (Morris, 2015). The global market also shows us the need for cost-effective product development and a profitable end product. Cost-effective design of new products requires assessing local and global market conditions, manufacturing capability, and the future health of the global economy.

In this complex market environment, creating cost-effective products from ideas can be an extremely complex process because it requires an understanding of the marketplace, knowledge of product development, and capital investment. This book concentrates on bringing innovative products to the market. It is a how-to guide. You can use this book based on your needs: assessing your idea for manufacturing, creating cost-effective marketing strategies, securing legal rights to the idea, or simply building leadership.

The outline of this book is straightforward, as shown in figure 1. The book starts with an idea and takes the idea all the way to the final product in the marketplace. Chapter 1 discusses idea creation and evaluation of initial commercial suitability. Chapter 2 further evaluates the attractiveness of the idea in the marketplace and analyzes potential competitors to determine whether the idea is indeed viable in terms of technology and long-term profitability. If the product idea doesn't pass the checklists in chapter 2, I suggest you improve the product idea or create another idea. Chapter 3 outlines the steps to legally secure your idea. If you would like to license or manufacture the product on your own, those steps are outlined in chapter 4. Chapters 5 and 6 provide guidance for prototyping and low-cost manufacturing. Chapter 7 focuses on the marketing plan and the tools needed to get your product to the marketplace. Tips for advertising and sales strategies are in chapter 8. The appendix contains many useful tools to evaluate your idea and create a business plan, a checklist for funding events, and tips to develop and improve your personal skills.

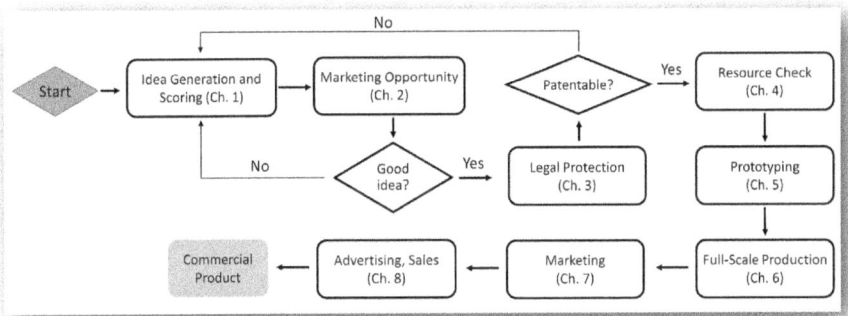

Figure 1. The outline of the chapters in this book

This book was written not by a business-school professor but rather by someone who has created a new product from just an idea and taken it to the marketplace. Without proper practical information, guidance, and a step-by-step approach, transforming an idea into a commercial product is nearly impossible. I believe ideas are only the beginning; developing a new product is like a climbing mountain. Without the proper skill set, guidance, and a little luck, it is possible that you will never get off the ground.

I have to admit that I find most business books boring. They focus too much on theory or fundamentals of design or marketing, and most of them are not based on real-life experience with product commercialization. When I was in college, my favorite class was with a professor who would talk about his real-life experience with the topic rather than recite facts from a book. I still remember his classes because I found his style to be very effective.

I have used information boxes that contain helpful tips throughout the book. If you are looking for particular information quickly, you can use these information areas. The information areas give summaries to capture the main ideas and explain the most important tips that I learned when developing my product.

Also, a wide variety of appendices and other resources for the book are available online. These resources include data, templates, links to suppliers, and lists of publications. The resources may be accessed via www.trainova.com.

The first chapter will cover innovation basics, idea generation, and scoring techniques.

CHAPTER 1

IDEA

In This Chapter

- Describing invention versus innovation
- Identifying needs for invention
- Creating ideas to address customer needs
- Scoring ideas as commercial products

Changing our direction according to the wind truly matters for those who want to create something new, as in my invention story. When I was younger, I worked as a machinist in a small town in Turkey. Like many young people, I wanted to see the world and meet people from other backgrounds. The easiest way to do this was to go to college. I studied engineering and moved to the United States for graduate school. I found that I enjoyed doing research, and eventually I worked as a scientist in a national laboratory while getting my PhD. After graduation, I worked as an engineer in several different fields, including the aerospace, automotive, and electronics industries.

Over the years, I have had the opportunity to work with and meet many incredibly talented people from a variety of disciplines and backgrounds. I found that everyone has a different way of thinking, working, and looking at life. Recognizing these differences has helped me to learn more about other cultures and disciplines. In every project, my hands-on experience in manufacturing helped me to create something new and to collaborate with others by contributing to their projects. However, I never thought of inventing something or considered myself an inventive person until starting to work as an engineer.

> There are clearly different outlooks among operations in the workplace. A machinist finishes the work that needs to be done according to the drawings or customer instructions. A scientist uses science to explain the existing issues and problems. An engineer uses science to create technology and new products. It is important to have all these capabilities when dealing with new technologies and applying them to new products.

My first engineering job after finishing my PhD was in the aerospace industry. My colleagues were very innovative, and they enjoyed using scientific principles to create new products and develop new applications. One of my colleagues mentioned an invention that he had developed in his personal time. I was fascinated to hear about it, and he inspired me to create something new using the skills that I had acquired during my years of doing research.

My colleague got me thinking about what I would like to do in my personal time. My wife and I had just purchased a house with a work space, and I realized that I wanted to create useful gadgets for others. This was my chance. With my interest and experience in machining, I first tried to purchase machining equipment, such as a lathe, milling machine, jigsaw, and so on, to build prototypes for my ideas. Unfortunately, these types of industrial machines are large, heavy, and too noisy for a home garage.

The idea of having a full-scale machine shop in my house didn't work out, but it made me think like my inventor friend. Instead of using industrial machines, I started thinking small—what small tools and equipment did I need to make projects easier? I found myself thinking about a template with which I could drill a hole an accurate distance from a material's edge. Then, after drilling the materials, I could assemble them using screws, rivets, or nails through the drilled pilot holes. The tool had to be simple to use and operate, not to mention affordable. This is where the idea of Measudrill started.

I first checked the available tools on the market. There were no tools with functions, costs, and dimensions similar to this measuring and drilling template idea. I thought there had to be other people like me who wanted to create things on their own as a hobby, who liked to own useful gadgets. I also thought there had to be enough people in the marketplace who would be interested in my product idea.

The idea of creating something new was exciting to me. It was just an idea, though. I had to keep improving the idea and making various prototypes. It is a true journey to fully develop an idea without knowing whether it will be commercially successful. As an engineer, I had never spent much time thinking about creating something new for others. Was it an invention or an innovation? I kept thinking about the differences between them.

1.1. Invention versus Innovation

"Invention" and "innovation" might be two of the words most overused by companies in the marketplace. Many companies use the words "invention" and "innovation" to boost the image of their products and lead to better competitiveness in the marketplace. As the needs of society change over time, constant adjustment and improvement of current products have become essential. Western firms spend $550 billion per year on research and development (R&D) alone to find out the needs of customers and come up with the solutions for those new needs (Radjou, Prabhu, Ahuja, & Roberts, 2012). With so much money being spent on R&D, it is crucial to know what differs between innovation and invention.

By a natural, internal driving force, we have been inclined to constantly improve and transform our lives. In this regard, the improvement of society has taken two distinctive paths: **invention**, in which we create new solutions to fulfill our needs, and **innovation**, in which we find ways to improve an existing solution to better meet our needs. Innovation is the constant refinement of an invention. Most inventions go through some innovation to fulfill a practical application. Inventors need to be equipped with the proper skill sets to transform ideas to products for the customer. It is also important to know what kinds of inventions they can deliver to the marketplace and the end customer.

1.2. Types of Inventions

When we hear about invention, we usually think of a new physical product. This is a limiting description of invention. The end result of invention can be anything, including but not limited to a physical product, an intangible product, or a service. There are many types of inventions:

- **Product invention:** a new or improved product. For example, a new portable device called Nima by 6SensorLab can detect gluten in food at levels as low as twenty parts per million, and it thereby allows for stress-free dining for people with gluten sensitivities (Robins, 2015).
- **Process invention:** a new set of operations to achieve a better result—for example, a new joining process called friction-stir welding developed by TWI Institute. In traditional joining processes, most metals require a filler metal and shielding gas, but friction-stir welding uses friction to join metals without need for a filler metal or shielding gas.
- **Organizational invention**: a new type of organization to do things—for example, a new venture division, internal communication system, or accounting procedure.
- **Management invention:** a new way of managing a business and operation. For example, the TQM (total quality management) system is a comprehensive and structural approach to organizational management that seeks to improve the quality of products and services through ongoing refinements in response to continuous feedback.
- **Production invention:** a new way of producing goods, such as new production planning software to reduce the waiting time between operations or a new inspection system to automate the formerly manual cosmetic inspection of a product.

- **Commercial/marketing invention:** a new way of entering a product into the market and creating new sales strategies. Most companies use these strategies to attract more customers and boost their product sales.
- **Service invention:** a new or improved way of providing services. For example, Internet-based insurance services reduce the number of local agents, reducing the cost of insurance. Therefore, they are more attractive to customers.

It is important to emphasize that invention should address the unmet needs of the customer. It should solve a problem that the customer faces. The magnitude of the problem solved determines the importance of the invention. When it solves a problem, the end result of the invention process can be an improvement in several aspects of the invention, such as function, cost, usability, quality, or environmental friendliness. For example, a newly developed lightweight cell phone technology can increase usability, and customers benefit by carrying the least weight.

Some inventions have a positive result but also have negative effects in other areas. For example, Zyklon B was originally developed in the early 1920s as a pesticide and disinfectant. Later, it was used as a chemical weapon in World War I because it had hydrogen cyanide as an active ingredient (Hanke, 2013). It is good idea to prevent misuse of an invention by thinking about other possible uses.

As we discussed, innovation is a process of improving existing products or inventions. Identifying new technologies and adapting them to current products are crucial to the success of an innovation. In the current marketplace, innovating existing products is a highly competitive area; we have lightweight cell phones, nonstick coating in ketchup bottles, computers with more capabilities, and so on. However, there are also newly invented products that address new needs of society. For example, a new moldable rubber glue called Sugru can be shaped like Play-Doh and sticks to almost any material. It can be used to fix gadgets and broken parts, with an unusual sticking capability compared to adhesives, sealants, and glues (Morrissey, 2016).

In the next section, we will discuss the fundamentals and historical development of invention.

1.3. Why Do We Invent?

Invention is driven by the needs of the customer, the market demand, strategic opportunity, business needs, social needs, environmental considerations, customer requests, technological advances, and legal requirements. **Need** is the driving force of human nature. Our basic needs are historically centered on food, water, shelter, transportation, and sleep (Berkowitz, Kerin, Hartley, & Rudelius, 1999). **Wants** are formed from needs and based

Market Introduction of Brands
Schweppes (Drink: Need): 1798
Levi's (Clothing: Need): 1850
Coca-Cola (Drink: Need): 1898
Boeing (Airplane: Want): 1916
Adidas (Shoe: Need): 1920
McDonald's (Food: Need): 1937
Nike (Shoe: Need): 1972

on cultural preferences. For example, everyone in the world needs to drink water, but cultural preferences (wants) tell a person in the United States to drink Coca-Cola, a person in China to drink tea, and a person in the Caribbean to drink coconut water. Wants for some goods have evolved from urbanization. If you are in a small city, you may need a car in your daily life, but if you are living in a large city, you may need a car only if public transportation is not available. I remember when I came to the United States first. While I had never driven a car or needed one, I had to buy a car because most of the places I needed to go, such as school and the grocery store, were not close by.

When backed by buying power, wants become **demands**. For example, most people can buy a basic cell phone, but they prefer to buy smartphones, which are more expensive. The demand for luxury goods and cars has been rising globally. For example, an Hermès purse sold for $298,000 in 2016, and the brand is known for its multiyear waiting lists for purses. There are both the want and the buying power in the market to support this demand (Huen, 2016).

It's hard to believe, but not that long ago, the washing machine, refrigerator, dishwasher, vacuum cleaner, and cell phone were luxuries (or wants). Now they are essential workhorses (or needs) in our society. With globalization, the shopping behavior of humans changed in terms of needs and wants, because goods and services became more readily available for everyone. The constant improvement in the quality of our lives has always been a paramount desire of society. In the not-so-distant past, it was difficult to introduce new ideas/products, but with the introduction and mass implementation of the Internet, goods and ideas could move freely and quickly, and the time required to implement an invention changed dramatically.

If we look at products introduced in recent years, the speed of adaptation is overwhelming. For example, the iPhone was released in 2007 to the global consumer market. Currently, there are 2.6 billion smartphone users globally, and by 2020, there are expected to be 6.1 billion smartphone users (Lunden, 2015). Shopkins is another example of the fast growth of products. It is a collection of miniature characters of the everyday items that you can find in stores. Since the Australian company Moose Toys launched the Shopkins brand in mid-2014, more than six hundred million Shopkins have been sold in the United States alone (Berger, 2016) (Nechamkin, 2015). Such a massive market creation was achieved in a small time frame.

On the other hand, let's think about another technology: the electric car. Tesla created an electric car in 1931 and showed it to the media. The car was able to go ninety miles per hour and didn't require a battery. Maybe gas was too cheap at that time, but the idea was not successful. We didn't hear about the electric car again until recently, when gas became more expensive (Royer, 2012). In this example, Tesla was a very early adapter of new technology, but society wasn't ready at that time for electric cars.

The introduction and adaptation speed of inventions as commercial products have changed over time. We currently live in a market environment in which inventions can enter the marketplace much faster than at any time in history. In the medieval era, most inventions came from manufacturing defense products such as armor and swords. Products with advanced new technology were trade secrets, and the implementation of the same technology in another society took centuries. After the Industrial Revolution in the 1850s and the advancement of communication platforms, the time required for commercializing an invention decreased. The 1990s were when globalization took effect, and now we see new products in the market in less than six months due to fierce global competition. As an example, the first iPhone was released in 2007 and the fifth version in 2012: five versions within five years of introduction to the market.

1.4. Adaptation of an Invention

The adaptation of a new technology or product depends on the industry; adopting new technologies may take a considerable amount of time. Some industries require rigorous testing and validation before releasing a product to the market. The military, aerospace, and oil/gas industries are considered to be in this category because they consider the design life of their products to be more than ten years; reliability becomes essential, and additional verification is required for safe operation of the products. Some industries also have strong dominance in the market and don't need to introduce new products, because there is no real competition. Car companies can be considered in this category. If they dominate the market in a specific country, they may only implement very small changes in their products. For example, French auto company Renault manufactured the Renault 12 model from 1969 to 2000 without any significant changes (Renault 12, 2016).

Some industries require high-speed development of new products owing to constantly changing market conditions. Consumer markets, information technology, bio/life sciences, medicine, advanced manufacturing, and value-added agriculture are the leading fast adapters for new technologies because they have the highest rates of competition owing to large market potential. Massive seasonal campaigns resulting from increased competition among distributors and retailers drive down product development times.

The product-development process may be simple or complex. Depending on product complexity, a careful assessment of financial capital and resources required is needed to determine the commercial viability of the product. If an idea has commercial viability, R&D resources are needed to prepare the product for the marketplace. Table 1.1 shows the typical R&D funding and personnel needs for the successful launch of a product in different industries.

Table 1.1. Typical minimum investment required for new-product development in different industries

Industry	Investment needed	Resources needed
Energy	$1M	10+ people
Home market	$20K	3+ people
Aerospace	$10M–$200M	100+ people
Agriculture	$5M–$100M	20+ people

It can take several years for some industries to adopt new technology. For individual inventors, suggested industries for inventions are the home, electronics, and consumer markets. Innovational products in these industries/markets are more quickly accepted than the ones in military, aerospace, and agricultural industries. Inventions also require more capital and resources than innovational products do.

After reviewing the fundamentals of an invention, the next step is to generate an early idea for a product.

1.5. Idea Generation

Where does an idea for a product come from? It starts with a customer analysis to understand what people buy and what they would like to buy. Customers are the major determinants for the success of a product in the marketplace; they may not buy the product if they don't feel it is right for them. Uncovering the unmet

> Ideas can strike us anywhere. Whenever you have a new, noteworthy idea, I suggest you write it down someplace so that you can revisit and improve the idea later.

needs of the customer is the key to a product idea and to product development. At this stage, you may not know what you will find from a customer analysis; the important thing is to keep listening to the customer and recording the information. The following steps are important to learn about customer behavior and environment:

- **Information gathering:** The only way to sell a product to a customer is to find out what the customer is buying and what his or her unmet needs are. You can do customer interviews and read reviews about current, similar products in the marketplace. With this information, you can identify the most common issues that the customer is experiencing with the current product. If you gather information about new technologies, such as cloud technology and the Internet of things, you may also want to review the available product information on the Internet to learn about customer expectations. You can use the template in appendix 1 to capture your findings. After gathering your findings on the customer, categorize the customer information into the following categories:
  - **Who** is the customer? Age, gender, income, geographic location, and loyalty to the business/brand.
  - **What** does the customer buy the product as? Need (groceries, diapers, etc.), want (toys, gadgets, etc.), or desire (expensive cars, for example).
  - **Why** does the customer buy the product? Personal use, training, gift, donation, and so on.
  - **When** does the customer buy the product? Holiday, seasonally, year-round, daily, and so on.
  - **How** does the customer buy the product? Online, big box store, direct mail, and so on.
- **Purchase evaluation:** If you are focusing on a particular product, you should check whether the product fits the customer's needs or wants. The following questions are useful to create a dialogue with the customer:
  - Does the product satisfy needs or wants? Areas for review include price, quality, availability, expectation, delivery, buying experience, and so on.

- What is the ideal customer experience for buying and using the product?
- Is there a gap between the customer's actual and ideal experiences?
- What are customer beliefs about and associations with the buying process?
- What barriers prevent some/all potential customers from buying the product?
- What opportunities are there to enhance the customer's experience?
- **Defining unmet and underserved needs:**
  - What are the areas in which the current product does not fulfill the customer's needs?
  - Are there any opportunities to improve the product for function, price, or usability? Identify the target customer and the opportunities.
  - Evaluate the attractiveness of the opportunities.
- **Report:** After collecting the information, organize and analyze it; the information may lead to a new idea, product, or solution. Appendix 2 shows an example of an idea-creation template.

Idea generation is not about finding one right idea. You need to create as many ideas as you can. If idea-generation activities are done with a team, the team leader is expected to encourage more ideas and ensure that team members feel welcome to provide various ideas about a solution. Idea-generating sessions without enough brainstorming activities can lead to a limited number of ideas. This type of approach can lead the group to a solution in which customer needs are not addressed.

After creating as many ideas as you can, the next step is to score them based on current needs and market conditions.

1.6. Idea Scoring

During idea scoring, there is additional brainstorming, creation of more ideas, and refinement of ideas, which can lead to a better product. Don't eliminate ideas immediately. An idea doesn't need to be perfect; it can be tweaked as it evolves. When developing a new or improved solution to an existing problem, it is wise to ask the following questions:

- **Function:** Are the functions of the concept well defined?
- **Target:** Is there any target industry? Is there any target customer type? Are both the target customer type and industry stable or growing?
- **Manufacturing**: Is it easy to prototype and manufacture the concept? Are there any environmental or safety issues with manufacturing the concept?
- **Competitor**: Are there any similar products on the market? Is the product a duplicate invention or innovation? Are there any possible intellectual property issues with the new product?
- **Price:** Is the early sales target for the product well defined? Does the sales price sound affordable to the target customers? Would you buy the product at this target price? Is it easy to find a distributor for the product?
- **Resources:** Is there any cost estimation for developing (R&D), prototyping, manufacturing, and marketing the product using your own or third-party resources? Do you have enough resources? See chapter 4 to further analyze the financial needs for product development and the potential payback time.

If the answers to all the above are yes, congratulations: you may have a product that is a candidate for the marketplace. If the questions above don't produce positive answers for the ideas generated, you need to further refine the ideas or come up with one that is more attractive. This exercise will ensure that you work on the idea with the highest potential for commercial success.

Regarding the marketing opportunities for your concept, use the template in appendix 3 to evaluate the commercial viability of your idea. In the next chapter, you will further evaluate your idea and prepare it for the marketplace as a product attractive to customers and distributors.

CHAPTER 2

MARKETING OPPORTUNITY

*The aim of marketing is to know and understand the customer
so well the product or service fits him and sells itself.*
—PETER DRUCKER

In This Chapter

- Creating a market opportunity analysis
- Creating a competitor analysis
- Determining whether the product idea is an attractive opportunity

When launching a new product or one with new or improved features, the most difficult part for many companies is identifying and reaching the customer. Unfortunately, it's a bit of a chicken-and-egg situation. Customers need to know there is a product before they can buy it, but most big box retailers want to know their customers will buy a product before they will carry it. Getting a new product to market can be tricky without a proper analysis. A marketing opportunity analysis can help you identify the potential magnitude of the marketing success of a product idea.

2.1. Marketing Opportunity Analysis

The commercial success of a product in the marketplace is highly dependent on the market conditions. Careful analysis of the market can help mitigate issues during product launch and sales. In other words, if the product has no chance to sell on the market, you can determine that at this stage rather than later. A marketing opportunity analysis will help you identify strong competitors in the marketplace and distinguish your product from others.

Before you focus on the marketing opportunity analysis, discovering what consumers want is key, as customer behavior controls the market. The **market** can be defined as the group of people with the desire and ability to buy a specific product. **Marketing** is the process of planning and executing the conception, pricing, promotion, and distribution of ideas, goods, and services to create exchanges that satisfy individual and organizational objectives (Berkowitz, Kerin, Hartley, & Rudelius, 1999). The following are needed for marketing to occur:

- Customer: two or more parties (individuals or organizations) with unsatisfied needs.
- Product: an item, service, or process to satisfy the customer's needs.
- Price: the amount of money that is exchanged for the product. Ideally, the price should satisfy both buyer and seller.
- Place: the location for customers to obtain a product. For new products, placement is not easy. The customer needs to know that the product exists and where to buy it, but big box stores often do not stock a product unless there is a proven customer demand.
- Promotion: communication between seller and buyer to raise awareness of a product or brand (Berkowitz, Kerin, Hartley, & Rudelius, 1999). Promotions encourage the customer to buy the product much faster. Discounts, coupons, and seasonal product campaigns are examples.

The successful launch of a new product highly depends on how well the product addresses the customer's needs and expectations. In fact, we see successful products and buy them every day. The majority of new products, however, fail, and the failed products fade from public sight and disappear. For example, floppy disks, dial-up modems, dot matrix printers, and zip drives are some of the computer technologies that were very popular once and are now obsolete. Reasons for new-product failures include the following (Urban & Hauser, 1990):

- Too small of a target market. Either the market is not ready, or there are not enough customers with buying power.
- Insignificant point of difference with current product portfolio on the market. There is no difference between the product and its competitors.
- Poor product quality. Customers don't like the product and leave negative comments in social media or on the retailer's website about the product.
- Price strategy is unknown. Target customers can't afford the product, and product price isn't adjusted to them.
- No access to customers. The product is offered in very limited locations.
- Bad timing. The product was introduced to the market either too early or too late.
- Poor execution of the marketing mix. Customer is not sure what the product does. The advertising materials are not clear.

Two of the main struggles for most companies are to identify the right customers and to meet their expectations. To understand customer expectations is to identify the real intended customer. Nearly all companies use marketing segmentation to identify the customer for the product. **Marketing segmentation** is a crucial part of marketing: you need to know who the primary customer is. By refining the list of the potential customers in terms of demographics, you can better target their needs and expectations.

The marketing segmentations for Measudrill are as follows:

- Who: someone who is creative and likes hobby projects and uses gadgets to fix and create things
- Occupation: homeowners, DIY hobbyists, carpenters, machinists
- Why: to not have to buy and carry around expensive machines

If the product enters the market at the right time, this helps product performance and long-term profitability. For example, toys shouldn't be launched right after the holidays, because the demand is minimal at that time. Before a new product enters the market, its success probability needs to be determined, and the product needs to be evaluated to see whether it is an attractive opportunity. This analysis is called a marketing opportunity analysis, and without it, most products are destined to fail. A **marketing opportunity analysis** is a tool to identify and assess the attractiveness of a business opportunity. In terms of bringing value to customers, here are the considerations:

- Trapped value: values expected to be stable in terms of efficiency, accessibility, and customer empowerment
- New value: values considered to be new features in terms of personalization, extension, community building, and collaboration
- Horizontal plays: stable processes and operations to improve functional operations
- Vertical plays: new processes and operations to improve industry-specific business activities

In terms of features, a market opportunity analysis uses the following key elements (Rayport, Jaworski, & Breakaway Solutions Inc., 2003):

- Competition: Analyze the competition environment and market. Find the main competitors and determine their key strengths in the market.

- Company: If the analysis is done for the company rather than the product, analyze the current state and resources of the company. If the analysis is done for the individual inventor, use section 4.5, "Estimating the Cost of Making Your Product," to evaluate the resources needed to commercialize the product idea.
- Technology: Analyze the readiness of the technology or other alternative technologies delivered with the product. If the product uses stable and well-known manufacturing processes, you can skip this section.
- Market attractiveness: Assess whether the product idea is an attractive opportunity in terms of marketing segment, long-term profitability, market size, and growth rate.

2.2. Competition Analysis

Without knowing the major players in the market and identifying their strong and weak points, you have little chance of launching a new product successfully. A **competitor analysis** is the first step in the marketing opportunity analysis to identify the major competitors in the market. Competitors are analyzed in terms of product and service selections.

Two types of competitors are generally considered for competition analysis: direct and indirect competitors. Direct competitors offer similar products or services. Analyzing direct competitors may not be enough because other companies in different industries may provide products that can replicate your product idea. Indirect competitors are companies in different industries offering products or services with the same function as substitute producers. Table 2.1 shows competitor manufacturers for Measudrill. The table lists tool manufacturers that may launch similar products in the near future. Because Measudrill is a new type of product, no differentiations were made between direct and indirect competitors.

Table 2.1. Competitor manufacturers for Measudrill (Tool, 2016)

Manufacturers	Brands	Products
ABM Tools	ABM	Industrial tools, cutting tools, measuring and marking tools, gauges, magnetic and dressing tools, tool holders, punches, vices, and clamps
Ideal Industries	Western Forge, Pratt-Read, SK Hand Tools	Hand tools
Lie-Nielsen Toolworks	Lie-Nielsen	Planes, other hand tools
Stanley Black & Decker	Black & Decker, DeVilbiss Air Power, DeWalt, Porter-Cable, Bostitch, Mac Tools, Proto, Blackhawk, Sidchrome, Husky	Handheld power tools, table saws, stationary tools (Delta)
TTI	Milwaukee, AEG/RIDGID, Ryobi, Homelite, Hoover, Dirt Devil, Vax, Craftsman, Bissell, Dirt Devil	Numerous tools
TTS Tooltechnic Systems	Festool	Handheld power tools
United Pacific Industries Ltd.	Spear & Jackson	Gardening and hand tools

By completing a review of competitors in the marketplace, you can also further refine the competitor analysis. This assists with assessing competition intensity by identifying the underserved and most competitive areas, current competitors' strengths and collaborators, and competitive hurdles.

Because Measudrill doesn't have a direct competitor, further evaluation for the competitor is not possible. To show an example competitive analysis, table 2.2 shows the major competitors of home improvement retailers (Rayport, Jaworski, & Breakaway Solutions Inc., 2003).

Table 2.2. Market segment examples

Target Segments	Harbor Freight	Northern Tool	Home Depot
Cost conscious	+ Discounts, special offers + Easy return, exchange, and cancellation policies + Pricing, free merchandise, coupons	O High-end products with lifetime warranty	+ Location of store + Product selection and availability + Friendly and helpful personnel
Middle-income families with children	+ Tools offered at very low prices - No rental services for equipment	- Unavailability of garden tools, strict return policy, limited number of rental equipment services	O Variety of home improvement and garden tools, rental equipment services
High income / tech savvy	- No selection of high-end products, limited warranty of products - Poor location choices	+ High-end tools for contractors O Limited support from personnel	+ High-end products with lifetime warranties + Good locations and friendly personnel O Limited selection of high-end products

+ high performance level, O medium performance level, - low performance level

After checking whether the product provides unique features and benefits compared to the competitor's products, you can further evaluate whether the product idea is an attractive opportunity in the marketplace.

2.3. Opportunity Analysis

Identifying potential competitors and analyzing their products may be helpful for product success. However, competitor analysis alone is not sufficient to define market conditions. The next step is to create an opportunity story to define the opportunities in the market (Rayport, Jaworski, & Breakaway Solutions Inc., 2003). To create the opportunity story, you need to complete the following steps:

- Describe the **target segment**(s) within the selected value system.
- Articulate the high-level **value proposition**. What benefits will the product provide?
- Outline the expected elements of **customer benefits**.
- Identify the critical capabilities and resources needed to deliver the customer benefits.
- Lay out the critical "**reasons to believe**" that the identified capabilities and resources will be a source of relative advantage over the competition.
- Categorize the **critical capabilities** (and supporting resources) as in house, build, buy, or collaborate.
- Describe how the company will capture the value that it creates for its customers.
- Provide an initial sense of the magnitude of the financial opportunity for the company.

Here is the opportunity story for the Measudrill:

- **Target segment:** DIY, woodworkers, homeowners, handymen, and small business owners
- **Value proposition:** Precision drilling for any materials, creating templates, and creating gadgets at a low cost
- **Customer benefits:** Low-cost gadget and fixture making and enhancing customer's creativity
- **Critical resources:** In-house manufacturing; innovative, low-cost manufacturing processes; and handling the operations

- **Reasons to believe:** Local manufacturing; lower cost–based, locally operated design; effective design changes based on customer's feedback; and environmentally friendly design
- **Resource sourcing:** Local supply chain
- **How to monetize:** Target homeowners and fixers for advertising success
- **Opportunity:** Medium—early starter, strong trademark and patent, and a very competitive design compared to outsourced manufacturing

In the opportunity story, target segments were identified from the intended customer types. Value proposition was derived from the function and intended usage of the product. Critical capabilities were defined from the product development, manufacturing, and sales operations. The opportunities for new products were defined using the local manufacturing concept, which uses innovative, low-cost manufacturing processes.

2.4. Market Opportunity's Attractiveness

When the product is launched in the marketplace, competitors will evaluate it and respond with similar products if the new product is a success. The attractiveness of the opportunity depends on its long-term profitability and relative competitiveness. Before going too much further with product development, it is good idea to determine the magnitude and character of the opportunity. In the evaluation process (Rayport, Jaworski, & Breakaway Solutions Inc., 2003), the following criteria can be used:

- Level of unmet need and the magnitude of unconstrained opportunity
- Level of interaction among major customer segments
- Likely rate of growth
- Size/volume of the market
- Level of profitability

Here are five attractiveness factors for Measudrill analyzed more deeply:

- Unconstrained opportunity: Measudrill is a new product.
- Segment interaction: Measudrill is used in both machining and carpentry tools.
- Growth rate: Measudrill is expected to have 30–50 percent annual growth of customer market.
- Market size: The power tool market alone generates $26 billion in sales (Wright, 2015). Global power tool demand will rise 4.8 percent per year through 2018 to $32.9 billion.
- Profitability: There is a high profit margin for tools. The typical profit margins of various goods will be discussed and calculated in more detail in chapter 6.

Pulling all the collected information together provides you with the marketing opportunity analysis. In terms of assessing the opportunity's attractiveness, you can use competitive vulnerability, magnitude of unmet needs, interaction among segments, likely role of growth, technology vulnerability, market size, and level of profitability. Based on these parameters, here are the evaluation results of the market opportunity analysis for Measudrill:

- Competitive vulnerability: Negative
- Magnitude of unmet needs: Positive
- Interaction among segments: Positive
- Likely role of growth: Positive
- Technology vulnerability: Neutral
- Market size: Medium
- Level of profitability: Neutral

In the opportunity analysis, you can determine which parameters are the most important for you. For Measudrill, I identified the magnitude of unmet needs, interaction among segments, and likely role of growth as the dominant factors. You can evaluate your product idea based on your preferred factors and decide whether your idea has a good market opportunity. Once you determine that the product has significant marketing attractiveness, you can further evaluate whether it needs any legal protection to secure it in the marketplace. Use the template in appendix 4 to evaluate the idea as a commercial product. If the product is not attractive in the commercial market, I strongly suggest you find out the reasons and further improve its attractiveness. Also, the profitability of the product was not discussed because the manufacturing processes will determine the product cost. Those factors are discussed in chapter 6, "Costing for Full-Scale Production."

CHAPTER 3

LEGAL PROTECTION OF THE IDEA

*Ethics is knowing the difference between what you
have a right to do and what is right to do.*
—POTTER STEWART

In This Chapter

- Protecting intellectual property
- Filing patent and trademark applications
- Keeping intellectual property a trade secret
- Learning what to do if your invention is copied without your consent

Since new product ideas are very easy to copy in the commercial market, legally securing them is as important as creating them. Products should not violate the legal rights of other product owners in the marketplace or the intellectual rights of other idea owners. Preliminary assessment of the product idea prevents such legal violations of other inventors' intellectual property and helps better protect the idea in the commercial marketplace.

When you are checking the legal protection of an idea, it is important to work with someone who is experienced with intellectual property laws. Many patent attorneys offer consulting and assistance in creating legal documents and legally securing an idea.

Before you talk to any patent attorney, it is essential to take notes and archive related documents such as photos, videos, and sketches while you are creating the idea. If you create a prototype or pay someone to create a product or service for you, keep receipts of all transactions as a record. In case of conflict about the invention, you can use these notes and receipts to prove that you are the sole inventor and to verify the date of invention.

Once the invention process starts, it is a good idea to keep it confidential and limit discussions about the invention with anyone. If you get help from someone you know and trust, you can ask them to keep the information confidential, but I recommend having them sign a **nondisclosure agreement (NDA)**. There are many sample NDAs available on the Internet, and some of the templates are listed in the recommended reading list. To protect your idea, both parties need to sign the NDA before you disclose any information. When disclosing information, make sure that the agreement covers the other party not sharing information with third parties. The agreement does not usually prevent the signatories from using protected information to create a different product. If the product-development effort involves some joint venture activities, both parties can sign a **joint development agreement** to decide who will own the intellectual property created by the joint work.

You also need to be careful when you are disclosing intellectual property information outside the United States. If the information is highly confidential or related to services provided to government contracting firms, you may be at risk of disclosing something that is covered in the International Traffic in Arms Regulations (McVey, 2011). Check the contents of the regulations before disclosing anything. You may also want to consult with an attorney if you have any concerns.

Keep in mind that even if the idea, prototype, or product is not covered in International Traffic in Arms Regulations, that does not mean you are free to disclose any information about it. If you disclose too much information, the party to whom it was disclosed can use or share the information and, in the worst case, could use it to construct the unit and sell it. Therefore, it is important to disclose only information the parties need to know. Also, when asking for a quote from a vendor, you can ask for a quote for individual parts rather than the whole assembly to be on the safe side. This way, the contractor or vendor will not know all design details.

Once you contact a patent attorney, the legal phase of the invention starts. The patent attorney will discuss your idea with you, take notes about your invention, and decide how your invention can be protected under the following legal devices: patents, trademarks, copyrights, and trade secrets. The following section explains these legal devices, their differences, and how to protect your intellectual property in the marketplace.

3.1. Patents

A patent is a legal contract between the inventor or the owner of intellectual property and the government. If the patent is filed in the United States, the US government is responsible. If the patent is also filed internationally, the World Intellectual Property Organization covers the legal protection of the idea. To sell an invention in the global market, it is advisable to file patents in the United States, the European Union, and other related countries. After the patent is accepted, the inventor or the owner of the intellectual property has the exclusive right to make, use, sell, or offer the item for sale. When applying for the patent, the inventor/owner needs to disclose how to make and use the invention. The exclusive rights are granted for a limited time, depending on the type of the patent application: up to twenty years for utility and plant patents and fourteen years for design patents. While the patent application is being prepared, the inventor needs to check available technologies to ensure that the invention doesn't violate or infringe on other patented disclosures.

Patented inventions allow the government to keep track of available technologies and advance new ideas. Though information on how to make the invention is disclosed, the public cannot produce or sell the invention without an agreement from the inventor. Patents can be used by purchasing the full rights of the patent, by **licensing** the right of manufacturing, or through a **royalty**.

If you need to prepare a patent application under time or budget constraints, you can start with a provisional application with no need for claims, which are the legal definitions of the invention. This provisional application is valid for one year and needs to be followed by a nonprovisional patent with claims. Depending on the function of the idea or product, there are essentially three types of patents:

- Utility patents: for inventions relating to processes, machines, articles of manufacture, compositions of matter, and certain living organisms
- Design patents: for inventions relating to the ornamental design of articles of manufacture
- Plant patents: for inventions relating to reproduced plants

To obtain a patent, a patent application must be prepared and filed. *It is good idea to have a working prototype of the idea before applying for a patent.* However, this is not required. It is important to think about the function of the invention and focus on what needs to be protected when filing a patent. The patent application must include the following information:

- **Description**: a brief explanation of the invention
- **Detailed description**: a more detailed section on the invention, discussing the originality of the invention and how it differs from other products that have been granted patents
- **Claims**: the essential functions and legal definitions of the invention
- **Figures**: graphical descriptions and formal drawings of the invention
- **Tables**: tabulated results of the invention, if any
- **Cover letter**: description of the patent application and contact information for future correspondence
- **Oath or declaration**: statement attesting that the invention is original and the applicant is the first inventor of the invention and accepting all criminal penalties if the nonoriginality of the invention is proven

The claims are the most important part of the patent application. They legally define the invention and control the scope of protection afforded by the patent (Sturm, 2015). If the claims are written too narrowly, they create unnecessary limitations. In this case, others can easily design a new product around the claimed invention, avoiding infringement and the necessity of licensing the patent. The scope of the patent decreases with narrowly written claims. If the claims are written too broadly, then they can contain the claims of existing patented inventions, and the claims will be rejected by the patent office or considered invalid in potential patent infringement suits.

After the patent application is filed, it is assigned to a patent examiner who has experience in the area of the invention. The examiner searches available granted patents and determines whether the claims are original. Any publicly available information supporting the originality of the patent's claims is called prior art, and it is essential to collect enough evidence to differentiate your invention from previous inventions.

After gathering the necessary information, the patent examiner determines whether the invention satisfies the requirements for patentability. If the patent application is accepted, the patent office files a record for the patent. If the patent application is not accepted, the patent examiner prepares an official letter, called an **office action**, and sends it to the inventor or the inventor's attorney. After the inventor/attorney reads the official decision letter and collects the required information about the originality of the patent application, he or she prepares an official response letter and sends it to the patent office. The examiner reviews the response to decide whether the patent application can be reconsidered and reports the response to the inventor/attorney.

Responding to the official letter is very important for patent application. If no action is taken, the patent application will be abandoned. According to the patent office, 95 percent of patent applications are abandoned because the inventor fails to pursue further action. In most cases, if you receive a rejection from the patent office, the telephone number of the patent examiner is in the response letter. You can call the person listed directly and ask for explanations. When you are preparing the response, it helps to know which area of the patent needs to be revised and what steps are needed for a successful response to the office action. Additional information about sample response letters is included in the recommended reading section.

Once obtained and issued, the patent can be used to prevent competitors from using the invention and gaining an economic or technological advantage. Also, the owner of the patent can license or sell patent rights.

Patent attorneys can prepare the legal documents and represent you in negotiations. Selecting the right patent attorney may determine the success of the product. There are so many attorneys out there, so you need to research any attorney you may work with. Referrals from colleagues can be helpful. Make sure your lawyer represents you and your best interests well. He or she should keep track of the patent application process and immediately inform you about updates. I remember a friend who chose a very busy patent attorney. When an interested company tried to contact his attorney for patent negotiations and licensing, his lawyer was always busy or unresponsive, so the interested company walked away. Everyone is busy, as we know, but you need to make sure that you choose an attorney who is professional and can respond quickly, write a patent application in a short time, inform you about the patent process, and represent you during negotiations.

3.2. Trademarks

If you are going to sell the product in the marketplace, it may be a good idea to apply for a trademark. A trademark or service mark is any word, name, symbol, or device to identify or distinguish your goods or services from other goods or services. Registering a trademark with the US Patent and Trademark Office gives the registrant the protection of the mark and prohibits others from using the same or a confusingly similar trademark. Allowing others to use the same marks would cause customer confusion and allow them to unfairly capitalize on your product recognition.

Trademarks often boost the image of the business as a serious manufacturer or brand in the marketplace. They also give the consumer confidence based on the reliability, quality, and safety of the goods or services identified by the mark. Trademarks are protected by the government to prevent infringement and provide a safer marketplace. Lawsuits for trademark violations are common. Recently, Intel sued SeaIntel for trademark violation. Because the court found that "Intel" was the dominant part of SeaIntel's trademark and designation, the company was ordered to cease trademark usage and pay a penalty (Njord, 2016).

Selecting the trademark is the first step of preparing the application. To ensure that you have a registerable trademark name, you can search the current registered trademark names on the Internet. If you choose a name too similar to a registered trademark, it may be difficult to get a strong protection for the trademark application.

> I found that applying for the trademark was much easier than applying for the patent, and the response was much faster.

It is often difficult to obtain a trademark for geographic names; surnames; or common words, symbols, or devices. Also, some trademarks have been used in the commercial market for a long time and have a "secondary meaning." In these cases, the registration of the trademark can never be obtained because the trademark name will be falsely descriptive. In other words, the trademark applications can't be associated with a specific brand. For example, *app store, aspirin, cellophane,* and *thermos* are all generic names that are not protected by trademarks (Quirk, 2014). However, some descriptive names, such as Holiday Inn (for motels), Quaker Oats (for oatmeal), are descriptive names, and they were registerable as a trademark because they were distinctive (Sherman, 2016).

When preparing a trademark application, you list the goods or services provided in connection with your mark according to their international class of goods and services. Classes are the categories of the goods and services that can serve a similar purpose. Goods are classified in thirty-four classes, and services are classified in eleven classes. The international classes are shown on the US trademark website (Trademark Class Headings and Explanatory Notes, 2012). The more international classes for which goods and services are considered for trademark application, the higher the fees for trademark searching and the application process.

In a trademark application, you need to represent the actual use of the mark in commerce or your good-faith intent to use it. You need to use the trademark in the sale or offering for sale of goods or services. The demonstration of the trademark needs to show the immediate availability of the product with the trademark to the customer. If the product is not available and it is shown on the Internet or displayed in a store as unavailable or coming soon, there is not enough evidence for the intended use of the trademark. Also, the trademark should easily be identified on the product. Placing the trademark on the product itself, on product packaging, or in the product manual will work.

After applying, you are automatically assigned a number, and most communication in trademark applications is done electronically rather than by mail, as with patent applications. After review by the trademark office, if there are objections or missing evidence, you can present evidence to counter the refusal to register your mark. The response letter may ask for clarification of the identification of goods or services and ask you to provide additional information. An experienced trademark attorney can help you prepare a response. You can respond to the decision during the opposition period, which starts immediately after the decision and is usually thirty days.

A certificate of registration (use in commerce) or notice of allowance (an intent to use) is created with a successful trademark application, and you can use the trademark in the marketplace. To continue using your trademark, you need to pay a maintenance fee every five to six years, depending on the trademark application class. After five years, you are required to file a declaration of continued use to prove that you are still using the trademark. After that, you need to submit the declaration between the nine- and ten-year registration anniversaries and then every ten years. If there is no action, the trademark will be cancelled or expired.

After publication of the trademark, the trademark must be used in the commercial market to be maintained. Also, the trademark owner must prevent others from using it. If a trademark owner allows other businesses to use the trademark to identify similar goods, then the trademark may lose some or all of its capability of identifying the owner as the source of those goods. Furthermore, if the trademark in question is allowed to become the common generic name by which that product is known, then trademark protection may be lost. Meanwhile, there are trademarks used as generic names that are still protected. Examples include Adrenalin, Bubble Wrap, Clorox, Frisbee, Jacuzzi, Jet Ski, Jeep, Post-it, Q-tips, Scotch tape, Tupperware, and Xerox (Trademarks, 2016). To protect the trademark and preserve it, the trademark owner may seek a court order prohibiting the infringer from using the trademark and requiring the infringer to pay appropriate damages.

Finding a trademark for my invention was not difficult. In order to describe my precision drill guide, I thought of the primary function of my invention. The template was doing "measure" and "drill" functions. I combined them and came up with "Measudrill." I used Measudrill as the trademark name because it was a coined name, and I thought there could be strong protection because it had not been registered before.

3.3. Copyrights

Copyrighting is a legal device to protect original works of authorship created in a tangible form. Copyright can cover anything from literature to musicals, from pictures to any graphic, sound, or video recording. Only the owner of a copyrighted work has the exclusive right to (1) reproduce the copyrighted work; (2) prepare derivative works based upon the copyrighted work; (3) distribute copies of the work to the public by sale, rental, or lease; (4) perform the work; or (5) display the work. Under copyright law, in the case of an individual, these exclusive rights exist for the life of the author, plus an additional seventy years. Copyrights are found in everyday life: the music on the radio, the poster in the hallway, and the movie the kids are watching all have copyrights.

It is important to know that a copyright does not protect the idea represented by a work. Rather, the copyright serves only to protect the particular expression used by the author to convey the idea. For example, an individual would likely not infringe a copyrighted book that described a new method of valuing real estate by simply practicing those teachings. However, if that same individual retyped or photocopied the book and distributed copies of the book for sale without permission of the copyright owner, such acts would likely infringe the copyright.

Copyright also applies when someone uses a copyrighted music arrangement but changes the lyrics slightly. For example, in 2015, Robin Thicke and Pharrell Williams were ordered to pay $7.2 million to Marvin Gaye's estate over the song "Blurred Lines" because Robin Thicke noticeably copied Marvin Gaye's 1977 hit "Got to Give It Up" (Grow, 2015).

Copyright law protection covers intellectual property whether it is published or unpublished. If you want register your copyright, the website is http://copyright.gov/eco/. Unpublished works are protected from the moment the work becomes fixed in a tangible form. However, recent changes in the copyright law give protection to unpublished works *before* the publication process (Strauss, 2013). A work is generally considered unpublished when copies of the work have not yet been distributed to the public. Once published, copyright protection may be diminished unless the work has been published with an appropriately located copyright notice. Such a notice will typically include a copyright symbol, the year of first publication, and the name of the copyright owner (e.g., © 2007 John Doe). However, not including the copyright notice doesn't mean that the original work is not covered by the copyright.

Remember, a copyright owner may prevent others from copying the protected work without authorization. A copyright interest may be retained by the copyright owner, or it may be licensed, sold, or partially rented.

3.4. Trade Secret

A trade secret is defined as information that is related to one's trade or business and would provide a competitive advantage if known. Trade secrets cover unpatented inventions, formulas, patterns, customer lists, recipes, methods of doing business, processes, machines, communications, and other forms of commercial information.

Trade secrets are protected by state law, and every state has different laws covering trade secrets. For example, the formula for Coca-Cola is the best-kept trade secret in the world. In order to be protected by the state, the information must actually be a secret before the court will offer protection. The trade secret, therefore, needs to be safely guarded with high confidentiality, security, and restriction of access to others.

Often a patent is applied to a product, while trade secrets cover the full details of making the product. If a competitor does reverse engineering, finds the trade secret, and files a patent, it may be able to produce the product without patent infringement. Trade secret violation by theft is a serious crime. For example, two ex-Coke workers were sentenced to federal prison for eight years because they wanted to sell Coke's recipe to Pepsi. Pepsi informed Coca-Cola, and the workers were arrested before releasing the trade secret information (Dornin, 2007). Just like patents, trademarks, and copyrights, trade secrets can be sold or licensed.

There are also considerations if you decide to sell your invention to a company or work with companies or contractors. As a rule of thumb, you need to be careful when you are working with companies in a joint venture not to disclose any confidential information regarded as a trade secret. I see this many times. If you interview someone as a contractor or freelancer and the person talks about the work that he or she did for another company in great detail, be careful. Most likely, any disclosed information will be shared with others soon. Even with an NDA, you should limit the information that is shared. This applies to data security as well. After Edward Snowden copied secret information from an NSA database with a simple flash drive, many companies adopted computer policies in which employees have restricted access to copy data to an external drive (Lewis & McVeigh, 2013). The same applies to electronic communications. Once you send confidential data by e-mail, the information stays on the recipient's computer and e-mail server. Make sure that you don't send information that could be a trade secret in the future.

Intellectual property can be legally protected in many ways. It is important to discuss the details of intellectual property with a patent or trademark attorney to help determine what best fits your idea. You can check with the US Patent and Trademark Office for more information, or check the list of resources in the recommended reading list.

If someone uses your patent, copyright, or trademark without your permission, you need to defend your legal rights. Consult your attorney; he or she may recommend that you initiate a lawsuit against the party using your intellectual property. The initiation of the lawsuit is called **litigation**. Before filing the lawsuit, it is important to contact the infringing party, ask it to stop infringing your patent rights, and ask for damages for unauthorized usage of the invention. You can consult with your attorney if you witness any violation of your intellectual property or want to know how to prevent such violations.

After you determine that the product idea needs to be legally protected, you can choose the proper legal channel. For products requiring patents, I strongly suggest starting with a provisional patent application because it is quick and cost effective, and you don't need to write the claims, which are the most essential part of the invention disclosure. You will need to extend the provisional patent after one year or file a nonprovisional patent application as a continuation of the patent to protect the idea. After the product is commercialized, you can apply for a trademark after the first day the product is offered for sale to the public. After determining the legal protection channel, you can further check whether you will license the application to a third party or manufacture the product with your own resources, as outlined in chapter 4.

CHAPTER 4

LICENSE OR MAKE IT YOURSELF

> *There are no secrets to success. It is the result of*
> *preparation, hard work, and learning from failure.*
> —COLIN POWELL

In This Chapter

- Decide to either license your invention or manufacture it yourself
- Evaluate whether you are ready to launch a product or business
- Self-assess your strengths and weaknesses
- Learn how to form a team for innovation
- Learn how to raise money to make your product

After protecting your idea with a patent, you can either manufacture it yourself or try to sell it to companies via a licensing option. Licensing is the agreement to transfer the manufacturing or to sell rights of the idea to another party for some time or indefinitely (Ferrell & Hirt, 2000).

Prior to initiating the licensing process, it is crucial to protect your idea because intellectual property violations may arise. For example, the company to which you license the idea can take your sample or learn about your invention and reverse engineer a similar, new product without informing you. It is sad but true, as was seen in the movie *Flash of Genius* (2008), about the legal battle between the inventor of intermittent windshield wipers and the Ford Motor Company. Therefore, it is a good idea to disclose a limited amount information during the initial licensing negotiations. Also, it is not a good idea to disclose the full details of the patent application. If you disclose the patent application number, the company or third party can check the information on the patent website, if available, and then reverse engineer your invention. Careful consideration is important when dealing with large companies. Representation by a patent attorney is extremely useful.

For licensing options, you can look for potential licensing companies and collect information from their websites to evaluate their potential interest in your product. In the recommended reading list, you will find further tips for licensing and a list of companies that are interested in learning about new products and ideas. While most companies look for products rather than ideas, some companies may be interested in ideas based on the product fitness. When submitting to licensing companies, it is essential to disclose limited information so that they have to come back to you for additional information. At this stage, it may be sufficient to disclose what the product does, the benefits to customers, and any available competitor products and differentiating factors. If you disclose too much information, you may have given your invention away.

The decision to license or to make a product on your own is not easy, and it requires a review of your available resources and an assessment of your own skills. Manufacturing the product on your own can yield a better outcome in terms of profit or reputation. However, it also requires an additional skill set. You may be a great engineer, but are you also a great businessman? Running a company is not for everyone, and you can evaluate whether you have the resources and commitment to run a business and manufacture products in the next section.

4.1. Are You Ready to Manufacture Yourself?

Launching a product in the marketplace requires concept building, prototyping, resources for manufacturing, and marketing capability. I believe that there is more to be gained by manufacturing your own product, but you have to have the resources. If you think you want to manufacture yourself, you need to carefully consider the following questions:

- **Health:** Are you healthy? Do you have enough energy to do this? This is a marathon, not a sprint.
- **Patience and perseverance:** Do you have enough patience to see this through? What about if things are not going well? What do you plan to do at that time?
- **Money:** Realistically, you have to have at least one year of funding to survive. Most start-ups close down quickly because the cash flow projections were not realistic, or the start-up funding was insufficient to keep the business running. You can use chapter 6 to estimate manufacturing costs.
- **Team person:** A successful business cannot be run by just one person. You will need to work with other people as an organization, as well as with external vendors, suppliers, and customers. Having strong interpersonal skills is extremely important to keep the business alive and successful. This can be an issue for start-ups because they usually lack team-building effort or direction.
- **Manufacturing/business experience**: You have to have interest and experience in manufacturing as well as business skills.
- **Niche product/service:** You have to have a something that differentiates your product from those of competitors, and there has to be demand for the product. This differentiating factor has been reviewed as the marketing opportunity of the product in chapter 2.

> **Starting with a niche product is very important.** When I was eighteen, I used to work at a flower shop. Flower shops became very popular, and within six months there were five new flower shops on one block. Because there was not enough business to support all of them, most of them closed.

If you decide to manufacture the product with your own resources, you need to have the interpersonal skills that are required to effectively interact with investors, team workers, and customers. Appendix 5 has a survey to determine your personal strengths. Then you can further evaluate building leadership, an innovative organization, and teamwork, and you can explore ways to raise money for the business, as discussed in the next sections.

A few considerations are necessary if you want to manufacture products. Being an inventor and running a business from your garage may not give you enough credibility with large retailers and distributors. It is a good idea to open up a business with a credible name for your distributors and customers. Selecting a business name is the first step. The name should be descriptive and easy to remember. Especially in the United States, creating a brand from non-English words can be quite ineffective because customers can easily forget or mispronounce the business name. Therefore, most small business owners prefer English names, which are easier to remember. If you have difficulty finding names, you can combine names and use them as business names, such as Johnson and Johnson or Proctor & Gamble.

The business location is another important consideration. Other key factors, such as the availability of talented employees, transportation facilities, parking facilities, building description and conditions, licenses and permits, machinery, machinery layouts, space for future expansion, available tax credits, design, and supplies, need to be considered when selecting a manufacturing space (Broom, Longenecker, & Moore, 1983).

4.2. Building Leadership

For your product and organization to be successful, you will need to be an effective, accountable, and personable leader. Most successful leaders focus on issues and situations, and they build relationships with team members. They build the self-confidence and self-esteem of others. They work on creating and maintaining constructive relationships and take initiative to make things better. A strong leader should lead by example and think beyond the moment. For example, a few years ago, Toyota announced the recall of 2.3 million vehicles due to faulty brakes. Toyota CEO Jim Lentz created a live forum with furious customers and answered all their questions. Such behavior minimized the damage to Toyota's reputation during the massive recall (Stansberry, 2010).

If you decide to manufacture and sell your product yourself, you may need to develop some additional soft skills to work with a team. **Soft skills** are the interpersonal skills that we use every day to communicate and interact with other people. They can help you organize people and effectively work with them. In other words, soft skills help you make sure that your team works together and achieves the defined goals. These skills can also help build your team and your reputation as a trustworthy business owner. Unfortunately, schools rarely offer classes to develop soft skills. You might consider checking for workshops and training programs offering soft skill development.

I believe communication and active listening are key soft skills for success. Evaluating soft skills is not difficult; see how long you can carry on a conversation with others. Or observe the gestures of others when they are talking with you. Once you start paying attention to what others really say and listening during conversations, you have the starting point to improve your soft skills for better communication.

It is important to pay attention, listen, and understand whenever you are talking with someone. Ask questions as needed, but it is important to carry on conversations according to the pace of the other party. I would compare a conversation to playing a table tennis game. You can't slam the ball in the other person's face; rather, you need to be in sync with the person's speed and level of understanding. Many technical people fail at this point because they explain their ideas too fast or in a too-complicated manner. Recent research showed that only 2 percent of the world population has formal training in listening, and the average person derives 55 percent of a message and meaning from facial expressions, 38 percent from the way the message is delivered, and 7 percent from the actual words spoken (Piombino, 2013).

It is important to use examples when explaining complex and technical concepts to ensure the message is clear. With this, you can also outline the main point of the conversation. As you develop relationships within your organization and with vendors/suppliers, you will hopefully come to know them and their communication styles as well. Remember, we live in a global society; you need to be able to communicate appropriately. Resources on how to be a more persuasive speaker are outlined in the recommended reading list.

Many people also have personal considerations that they need to confront to be successful. Three of the most common are fear of failure, fear of success, and fear of public speaking (Gallo, 2013). These have the following characteristics:

- **Fear of failure:** Many people have a fear of failure. They tend not to try things that are out of their comfort zones. They may not want to try something new to avoid being disappointed or embarrassed if they are not successful.
- **Fear of success:** Many people fear being successful. Being successful will push them out of their comfort zones and into new situations. It is easier to remain with the status quo or in jobs where they are overqualified but can live in their comfort zones.
- **Fear of public speaking:** This was found to be the most dominant fear of most people (even greater than the fear of death). Many people with this fear are reserved and avoid speaking in groups.

There are resources available to build personal confidence. Some are included in the recommended reading list section. There are also groups that help with public speaking, like Toastmasters International. Public speaking can be overwhelming; even Warren Buffett had a public speaking fear when he was younger (Gallo, 2013). With practice and determination, you can improve your confidence in public speaking.

4.3. Building Teamwork

An individual is a source of ideas and entrepreneurial spirit. However, no successful business can exist without teamwork. As one of my colleagues once said, "Most companies make buildings, gadgets, and equipment, hire engineers, scientists, and PhDs to get the job done, but at the end, it is all about the people, how we work together and effectively communicate." The strength of an organization relies on the foundation of the expertness of each member and how they interact with one another.

A team is a group of people assigned specific tasks to achieve a goal. Recognizing the occupational backgrounds of team members is necessary because the team needs to be built based on the functional expertise of its members. It is important for the team builder to understand what kinds of skill sets are needed for the work to be completed. In terms of the functions of physical product development, I tend to categorize team members based on the strong skills that they can bring to the table. I use the following categories to identify the functional strengths of people:

- **Shop-floor smart:** Shop-floor smart people are creative with tools and gadgets and able to **manufacture** prototypes, tools, gadgets, and products. They tend to be skilled at using power tools, hand tools, and gadgets. They are hands-on people. In order to build prototypes, manufacture the product, and create development projects, it is good to have them on the team.
- **Computer smart:** Computer smart people are very familiar with computers, recent technology applications, and computer-aided design technologies. They are good with the latest computer technologies. These people are necessary for the team to create design development files, catalogues, and manuals. They can also be responsible for the electronic data security of the project files and documentations.
- **People smart**: "Personable" people are naturally good with other people regardless of their backgrounds. They tend to get along with anyone, can make small talk, and are good at making others feel comfortable. These people tend to be good at **marketing** and sales.
- **Culture smart:** Culturally smart people can work with people of different cultures; they can adapt easily and understand cultural differences, expectations, and operations. They are necessary for **international business.**
- **Book smart:** These are people who are introverted by nature. They can be great for bookkeeping, **accounting**, running the company, and internal organizing and planning.

In the manufacturing industry, it is essential to have at minimum a **shop-floor-smart person** to play a lead role in defining the functions of the product, making the prototypes, and designing the production system; a **people-smart person** to advertise, market, or sell the product and mediate the interactions between the company and its customers; and a **book-smart person** to keep the business running and to negotiate with distributors and customers.

For more complex product development, building teamwork requires soft skills and understanding people in terms of background, age, experience, and culture. I have worked with people from around the world, including the United States, Turkey, Germany, Spain, Canada, India, Pakistan, Korea, China, and Taiwan, and have found that everyone thinks very differently and has different expectations. For example, Western cultures focus more on project goals, while Eastern cultures tend to build relationships between coworkers first. Therefore, it is important to create a communication platform for team members to contribute and communicate for the best outcome, as well as to outline the big picture of goals and expectations for the product-development project.

In such a complex and competitive market environment, many inventors and start-up companies need to seek out shared services or contract manufacturing assistance to reduce product and operation costs. You can use such shared services on design, manufacturing, and marketing operations and work with experts to ensure the high quality of the product. If you don't have the budget to hire them, you could work with them through sweat equity: you offer some portion of the profit from the new product in return for their work. If you do hire external resources, it is a good idea to have an agreement to outline the following conditions:

- Required tasks and milestones.
- Decision making (roles and responsibilities).
- Capital distribution (finance): how the manufacturing, packaging, and sale costs will be covered.
- Profit distributions: how the profit will be shared among the parties.
- Dissolution: if either party decides to terminate the agreement, how the other party will be notified and how the remaining units and profits (or expenses) will be distributed between the parties. This section can also cover the extension terms for the continuation of the partnership if the agreement is for a short time.

Once the functional team is formed and product development starts, working together can be chaotic if the team members are fairly new to one another. Each member will have different expectations and working styles. Once the functional team is operational and some work is done, you can check the work progress. If the work outcome is not producing the expected results or you would like to improve the project deliverables, you can evaluate the teamwork in terms of efficiency, interaction, and achievements and come up with the team's stronger and weaker points.

As shown in table 4.1, a SWOT (strength, weakness, opportunities, threats) analysis can be very helpful to identify the areas to focus on. It is important to complete the SWOT analysis as honestly as you can. If you have difficulty finding weaknesses, you can always ask for help from team members or others outside the team. Spouses, parents, friends, trusted business advisors, and colleagues can help identify your strengths and weaknesses. You can discuss the results with the team members and create a report. The SWOT analysis report can lead to action items for the team and to project improvement.

Table 4.1. Questions for SWOT analysis

Strength: • What are we doing the best? • What advantages do we have? • What do others call our strength?	Weakness: • What don't we do well? • What do we need to work on? • What should we avoid doing at all?
Opportunities: • What resources aren't we using? • What's the environment right now? • Are there trends we can take advantage of?	Threats: • What are our current objectives? • What might hold us back? • Is there anything that could jeopardize our organization's or project's future?

4.4. Building Inventive Organization

This section will compare the inventive environment of individual inventors to R&D facilities of large companies. Recognizing outlook differences can help you be successful in the commercial market against large companies. If you are working as a small team, you can skip this section and move on to section 4.5, "Estimating the Cost of Making Your Product."

Many companies compete in the global market in terms of manufacturing, speed, cost, and effective customer interaction. Improving business operations at all levels is important to keep a competitive edge. If the customer market and competitors are not analyzed and forecasted properly, the consequences of market fluctuations can be detrimental to the company. For example, Atari Inc. was a major manufacturer in the video game market in the 1980s. In 1983, the company introduced a new video game called *E.T. the Extra-Terrestrial* and produced twelve million copies based on the strong sales of previous video games. The company underestimated the massive success of a competitor's games, *Space Invaders* and *Pac-Man.* Having five million unsold copies of *E.T.*, Atari found a remote location in New Mexico and buried them (Atari, 2016). Producing large numbers of a product without checking the new products launched by competitors cost Atari millions of dollars at that time.

Building and maintaining an inventive organization is not an easy task. Team players need to be in sync on the organization's goals. The same problem arises in most large corporations. Organizational gaps, communicational gaps, issues with the infrastructure, and lack of understanding of the project goals lead to a less innovative process (Janov, 1994). If small organizations can find ways to compete with large organizations, they can be much more productive in terms of delivering innovative products to the market.

There are several myths about invention. Most people think it is all about the ideas. I believe an idea cannot be converted to a new product without effective project management. Invention requires an idea, motivation from the idea creator, a process to achieve the desired outcome, and finally a leader and a project-management team to execute the plan. The team must divide tasks and milestones, manage a partnership between external collaborators if needed, and create a realistic deadline for milestones. Also, because the product needs to be profitable, the cost targets for manufacturing and for the final product need to be established before launching the product.

Launching a product in multiple countries, especially emerging countries, may lead to additional opportunities. A common thought is that in poor countries, there is an opportunity to sell only to wealthy people (Govindarajan & Trimble, 2012). This is often not true. If the product provides status and popularity, many people in emerging countries will try to buy it even if they cannot afford it. For example, eating at McDonald's is not very exciting for most adults in the United States because you can find one every ten miles. But when the first McDonald's restaurant was opened in Moscow, Russia, in 1990, thirty thousand Russians visited on the first day and waited in line to pay the equivalent of several days' wages for a burger and fries (Keppler, 2016).

For the local market, customization of the product may be required to address the expectations of the customers. Creating a team by recruiting local employees is essential to compete in the local market. In that case, the bulk of future customers, the potential for future growth, local manufacturing capabilities, the social infrastructure, and potential new competitors need to be evaluated.

Application of technology can also improve the speed and efficiency of invention projects. Technology is knowledge applied to products or production processes. Technological products require a sophisticated team formed by interdisciplinary functions. The team members must be seasoned experts in their areas to effectively achieve the product-development goals and milestones. After the idea is confirmed, the R&D team evaluates the options for the product. The manufacturing team creates the prototypes and creates design for manufacturing (DFM) and design for cost (DFC) studies to further reduce the product cost.

Once an inventive organization is developed, employees can adapt differently to the changes in the organization, which might include changing from flat to divisional organization, from individual contributors to self-managed teams, from manual to automatic operations, from decentralized to centralized operations, or from short-term to long-term focus (Janov, 1994). Such changes should be carefully planned to ensure the minimum impact on the team's performance. In terms of response to changes inside and outside the company, organizations can be categorized as reactive, responsive, or inventive.

- **Reactive:** Employees typically care about how changes will impact them when changes occur. Even though the system has issues, they are not communicated to management. In this case, most employees start to differentiate themselves from management ("us" and "them"). Whenever there is a new change, employees describe it as "they did this; they did that." Mostly, workers think that the new changes will not produce any positive result in the company's business.

- **Responsive:** Employees and management are focused on responses. There are tools available to voice concerns and ideas for improvement in the workplace. Management occasionally creates feedback surveys to learn about employees. It is often difficult to change anything or implement new strategies because the organization has a strong culture and prescribed way of doing things.

- **Inventive:** Employees and management focus on service to the customers. Support of the total organization is focused on the product and the development of customer support. All business operations are managed organically, and response time is much faster than in other types of organizations.

The team and company should always focus on the product and the value created for the customer. Such focus can reduce the product development time and cost. After discussing the inventive organization, the next step is to estimate the cost of running the company that makes your products.

4.5. Estimating the Cost of Making Your Product or Opening a Company

Money is the fuel for inventors and start-ups. Without money, it is impossible to bring an invention to the marketplace as a product. Most start-ups and inventors run out of money and fail because they do not accurately predict the cash requirement for the commercialization of the product idea.

The cheapest way to invent, produce, and manufacture is to run operations from your house and look for a contractor company or a skilled tradesperson who can do the manufacturing for you. In that case, it is still essential to know the product cost, profit, and sales price. Table 4.2 shows an example calculation for the product cost of a small injection-molded plastic part. The same table also outlines the potential revenues based on selling the product in three different channels: wholesale, e-tail, and direct sale. Based on the potential sales volume, you can calculate the revenues and potential payback period of the product. Calculating payback period is important in order to have a realistic approach for the investment you are seeking. In table 4.2, advertisement and other expenses are not calculated, to estimate the minimum cost for the product development.

Table 4.2. Example calculation of product cost and payback time

Expenses	Total
Tooling for injection molding	$2,161.00
Total piece price (5,000 pieces, $0.20 per piece)	$982.00
Transportation and customs (if any)	$800.00
Final packaging	$200.00
Total cost	$4,143.00
Per Piece cost	$0.83
Potential revenues	
Yearly wholesale (20% profit, $1 per piece price, 200-piece monthly sale)	$2,400.00
Yearly e-tail (e.g., Amazon, $1.80 per piece price, 100-piece monthly sale)	$2,160.00
Yearly direct sale ($3 per piece price, 100-piece monthly sale)	$3,600.00
Total potential revenues (yearly)	**$8,160.00**
Payback period	**6.1 months**

If you decide to open a company to handle product development and manufacturing operations, it is important that you have funding for at least one year of operation. Table 4.3 estimates the yearly overhead expenses for renting a commercial space and hiring one employee for administrative work. You can adjust these numbers based on your location. The below table does not cover the advertisement costs for the product. To estimate the advertising expenses, see chapter 8.

Table 4.3. Estimated yearly overhead expenses for operating a company in the United States

Expenses (approximate, yearly)	Total
Rent (10,000 sq. ft., 929 m²)	$60,000
Restoration, paint, interior design	$10,000
Furniture	$10,000
Electricity, water	$3,000
Administrator ($20/hour)	$32,000
Subtotal	$115,000
15% contingency	$17,250
Total	$132,250

For additional and more complex accounting calculations, including purchasing space or hiring employees, you may want to review the cost estimate with a certified public accountant or cost specialist.

4.6. Raising Money for Your Business

Funding your initial product operations can be challenging because the product is unknown and the risk for failure is high. Unless you have an angel investor, a person or company who can fund your business, you will have to provide start-up funding to run the business and operations.

Starting a business on a smaller scale is less risky financially than a large operation. For example, it is not advisable to buy a large commercial property when renting a location is economical. Renting equipment and hiring freelancers may also be more economical. In the case of hiring freelancers, they need to be compatible with the team's culture and project goals. There are many resources to fund your business, and most of them are based on borrowing money or granting some share of the company. Here are the options you may want to consider:

- **Use your savings or assets:** If you have savings, valuable assets that could be sold, or wealthy parents, that's great. You may already have the money you need. For example, Papa John's founder John Schnatter sold his 1971 Camaro Z/28 to purchase pizza equipment and began selling pizzas to customers out of a converted closet (Hardigree, 2009).

- **Apply for grants:** There are many grants available for small business owners. Some of them, especially Small Business Initiation Research (SBIR) grants, do not need to be repaid. (If you have to pay back the funding, it is considered a loan.) You will need to work with an experienced grant writer, as these grants are very competitive and hard to get. Also, some grants require additional checkpoints for the project after awarding it to ensure that the grant is used for the awarded purpose.

- **Use crowdfunding:** You can use crowdfunding websites to raise money for your business. You can pitch your idea by making a video and a campaign for raising money. Crowdfunding websites work on different principles (Anastasio, 2016). For example, Kickstarter.com is an all-or-nothing platform. (That means you don't get any money if you don't reach your goal.) Indiegogo.com and GoFundMe.com have "flex" funding (meaning you get everything you raise). With social media and social networking tools, some crowdfunding campaigns have been very successful.

- **Borrow money from your 401(k):** Some of the money in retirement accounts can be used to fund a business without a tax penalty. You need to contact a tax specialist for the terms and conditions.
- **Take out a loan:** If you decide to get a loan, possible options include a small business loan or a personal loan. (Small business loans have a lower rate of interest.) In either case, your personal financial history is very important for the loan application process because it is a picture of your personal financial condition to date. A complete personal financial history is a record of borrowings and repayments and an itemized listing of your personal assets and liabilities. It will list your sources of income, such as salary, personal assets, and investments (stocks, bonds, real estate, savings accounts). Your liabilities in the form of personal debts (installment credit payments, life insurance premiums, mortgage status, etc.) must also be listed in detail. The following are the step-by-step procedures for applying for a business loan:

A. For a new business
   1. Write a detailed description of the business to be established.
   2. Describe your experience and management capabilities.
   3. Prepare an estimate of how much money you and others have to invest in the business and how much you are requesting to borrow.
   4. Prepare a current financial statement listing all your personal assets and liabilities.
   5. Prepare a detailed projection of earnings anticipated for your first year in business.
   6. List collateral you can offer as security for the loan, including an estimate of the present market value of each item.
   7. Take all the above with you to the lender. Ask for a loan.

B. For an established business
   1. Prepare a current financial statement (balance sheet) listing all the assets and liabilities of the business; do not include personal items.
   2. Prepare an earnings statement (profit and loss) for the previous year and for the current period to the date of the balance sheet.

3. Prepare a current personal financial statement of the owner or each partner or stockholder owning 20 percent or more of the stock in the business.
4. List collateral to be offered as loan security, with your estimate of the present market value of each item.
5. State the amount of the loan requested and explain the exact purposes for which it will be used.
6. Take the above financial information with you to the lender. Ask for a loan.

Below are some banking terms that are useful to know during the loan process.

- **Balance sheet:** a current financial statement. It is a dollars-and-cents description of your business, existing or projected, that lists all your assets and liabilities.
- **Profit-and-loss statement:** a detailed earnings statement for the previous year (if you are in business). Existing businesses are also required to show a profit-and-loss statement for the year of the balance sheet.
- **Assets:** accounts payable (money customers owe you), inventory (stock or merchandise), equipment (furniture, fixtures, machinery, delivery trucks), and anything that can generate cash.
- **Liabilities**: accounts payable (money you owe to suppliers), plus all current costs of doing business (mortgage payments, insurance, taxes, salaries, utilities) and anything you have to pay out or will cost you money in running the business.
- **Cash flow projection:** a projection of the cash (checks, money orders, credit card payments, digital wallet payments (e.g., Apple Pay), payments made via PayPal, etc.) a business anticipates receiving and disbursing during a specific time period. The time period usually is a month. The anticipated cash flow should be sufficient to meet the cash requirements for the following month.

- **Collateral:** a favorite word in the banking community. It means property, stocks, bonds, savings accounts, life insurance, and current business assets—any or all of which may be held or assumed to ensure repayment of your loan.

4.7. Next Steps

After deciding that the product idea is worth pursuing and can be funded, create a plan for making a prototype and for the next steps in the production. You can first define a work breakdown structure to outline the steps, as shown in figure 4.1. A work breakdown structure is a useful tool to explain the steps to people from different backgrounds.

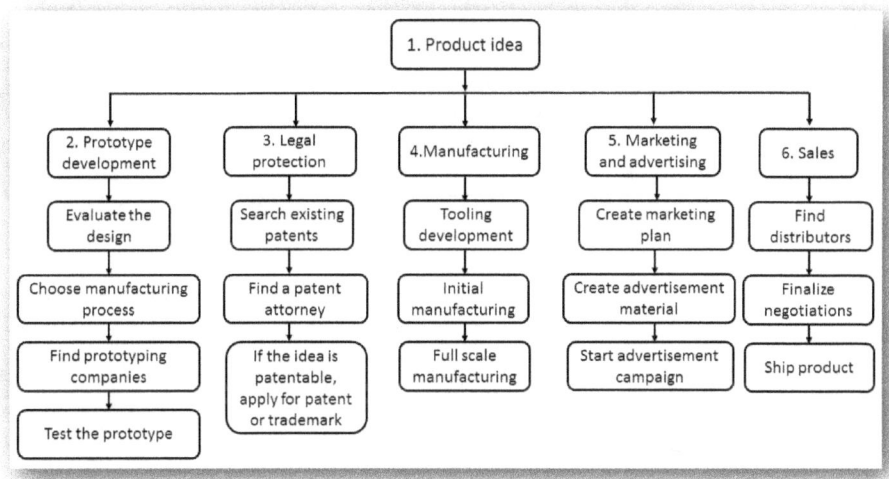

Figure 4.1. Work breakdown structure for product development

Any project without a timeline or schedule may not be successful. Because tasks can be interdependent and the project team needs to ensure tasks are completed on time, creating a schedule can make product development easier. A Gantt chart can outline the steps for prototype development. Table 4.4 shows an example of a Gantt chart to establish a timeline and schedule for required tasks. Most tasks on the schedule should be specific, and it should be possible to accomplish them in a trackable time period. One to two weeks is a good period within which to track progress. The time period for the individual tasks should also allow for unforeseen production delays and scheduling issues. Estimating such time delays and including them in the schedule can help reduce the number of iterations in the prototype development process and create a more flexible production process.

Table 4.4. Example project timeline schedule for prototype development

Task	Timeline					
	8/1/2016– 8/12/2016	8/15/2016– 8/26/2016	8/29/2016– 9/9/2016	9/12/2016– 9/23/2016	9/26/2016– 10/7/2016	10/10/2016– 10/21/2016
Make a prototype	▓					
Apply for patent	▓					
Find companies that can license the product		▓				
Negotiation starts				▓		
If no licensing company is found, make it on your own						▓

CHAPTER 5

PROTOTYPING AND LOW-SCALE PRODUCTION

The only source of knowledge is experience.
—ALBERT EINSTEIN

In This Chapter

- Prototyping
- Initial low-scale production
- How to keep the cost down

A prototype can be defined as "an approximation of the product along one or more dimensions of interest" (Ulrich & Eppinger, 2000). Prototyping is a crucial step in product development. For physical products, you should never invest money in a product idea if you cannot make a prototype. The prototype is an initial description of what you will provide to the market. If you are trying to raise funding for product commercialization, physical objects are also easier to show to others and use to explain how the product works. Prototyping also helps to refine the design and transform the idea into a product. The prototype can help potential customers and investors learn about the product and its functions, as well as identifying manufacturing and assembly constraints of the product before production. It is important to have a prototype when you discuss the idea with possible investors. I attended a funding event at which I showed my three best product ideas. One of them had a prototype, Measudrill. I showed investors the prototype and how it works. The funding committee members liked the prototype because it was more than an idea. They asked me to pursue it as a commercial product. After the meeting, I decided to focus on the Measudrill.

5.1. Prototype Development

Prototypes are manufactured in two types: **work-like** and **look-like** prototypes. In work-like prototypes, the focus is on function. For look-like prototypes, the look is more important, and the prototype may not provide any of the expected functions of the product.

> The best stores for raw materials, parts, and components for prototype building, based on availability, are McMaster-Carr, Grainger, Digikey, Online Metals, eBay, and Amazon.

Before producing a prototype, it is a good idea to plan what the prototype will do and how it will be produced. The plan is called the initial scale production plan, and it includes the following steps:

- Create a specification. This is the early definition of what the product has to do.
- Gather the materials and create a complete bill of materials. Make sure that you choose materials that are representative of the working prototype. This step eliminates various iterations in the prototype development.
- If you are outsourcing, find out who can manufacture the prototype. You can check with local shops or friends who have the special skills to manufacture prototypes.
- Create a budget for the prototyping.
- Create a financial plan to cover the cost of prototype manufacturing.
- Create a schedule for prototype manufacturing.

During product development, the complexity of the product defines what type of prototype to build and how to incorporate the lessons learned from prototypes into the large-scale product manufacturing plan. Physical prototypes are preferred for products with simple functions. Analytical prototypes are used for products with complex functions, such as those using electrical or mechanical principles. Creating an analytical (digital, computer aided) rather than a physical prototype has the following advantages:

- Analytical prototypes are much more flexible than physical prototypes because design changes can easily be incorporated.
- Analytical prototypes may reduce the risk of costly iterations.
- Analytical prototypes can expedite other development steps, such as manufacturability and assembly.
- Physical prototypes are required to detect any unanticipated issues, such as assembly and mating part compatibility. For example, Chrysler had an issue with assembling the PT Cruiser's engine in a moving line; the assembly of one engine took a whole day instead of a couple of minutes (Poeth, 2010). As a result, the company had to spend weeks to recover from this production problem.

Analytical prototyping has benefited from the use of 3-D printing technologies, in which the digital design can easily be converted to a physical product. 3-D printing constructs shapes layer by layer from a digital file (Krassenstein, 2015). After the design is complete, the design file is converted to a special format called stereolithography (STL). The design file is transferred to a 3-D printer, and the printer melts plastic materials and adds them layer by layer based on the coordinates of the shape.

Special materials such as ABS or PLA, which melt and bond to the previously printed layers easily, are usually used (Wilson, 2013). With the availability and reduced cost of 3-D printing systems, such techniques can be considered the future of prototyping. Currently, metals like certain steel grades can also be printed by using a powder bed and fusing it with lasers. Aluminum printing is possible; however, it requires a vacuum environment because the aluminum powder is combustible.

Figure 5.1 (a) shows the design and guides of Measudrill, produced with 3-D printing technology. The part was designed in free software called 123-D by Autodesk. The final shape is shown in figure 5.1 (b). The overall process is fast; it took about ten minutes to print this part for less than two dollars.

(a)   (b)

Figure 5.1. (a) Design for 3-D printing; (b) 3-D printed guides

During the early prototype-design process, a team meeting of functional groups is required to define specifications and performance expectations (Boothroyd, Dewhurst, & Boothroyd, 1994). This meeting helps to reduce the iterations from prototype to final product. The outcome of this meeting is outlined in the development document, and improvement opportunities are identified by the manufacturing team.

The meeting also confirms the suitability of the selected manufacturing processes for the product and their ability to produce quality products with the lowest cost. If the product will be manufactured by an outside vendor, it is important to do an internal audit to identify the weak points in the quality and in the manufacturing system of the vendor, especially if the vendor is expected to develop a manufacturing process specification for the product right after the prototype is made. If something goes wrong in terms of the product, the vendor needs to have a record to ensure that the failed unit can be traced back to the manufacturing date and that the root cause of the failure can be identified easily for preventive action. Of course, these are recommendations for high-quality products with massive production volume, and you may not need such planning in the initial stage of product development.

5.2. Prototype Manufacturing

To create a physical prototype with conventional manufacturing processes, it is more practical to use common raw materials such as metals, plastics, ceramics, glass, wood, and fabrics because they are economical and commercially available. Most economical and commercially available raw-material products are sold in common shapes such as bars, flat sheets, and U-channels.

Prototypes with complex geometry should be manufactured with a casting process, in which an external mold of the prototype is created and molten metal is poured to achieve the final shape. For structural prototypes, metals such as steel, stainless steel, and aluminum can be used. For medium-strength prototype applications such as toys, packaging, and electronics, plastic materials can be used. Wood and modeling compound can also be used in nonstructural applications. Table 5.1 shows the general, manufacturing, and assembly characteristics of various materials.

Table 5.1. Various materials used in building prototypes

Material	Advantages	Disadvantages	Machining/cutting	Joining methods
Steel	Cheap, easy to find (look for 1040, 1080 grades for plain carbon steel; 4140, 4340 for high-strength steel grades)	Heavy, needs painting for corrosion protection	Moderately easy to cut/machine	Spot welding, stick welding, riveting, bolted joints
Stainless steel	Very easy to form, bend (look 304, 316 grades for stainless steel; for strength look for 17-4PH stainless steel)	Very expensive	Easy to cut and machine except 17-4PH grade	Spot welding, stick welding, riveting, bolted joints, TIG/MIG welding
Aluminum	Light, very easy to cut, does not require painting (look for 6061-T6, 5052 grades for common alloys)	Moderately expensive	Very easy to cut/machine	Riveting, TIG/MIG welding
Wood	Inexpensive, easy to machine	Not good for impact loading	Very easy to cut/machine	Bolting, riveting
Modeling compound	Inexpensive, easy to shape	Not durable	Very easy to shape	NA
Nylon, polycarbonate, polypropylene, polyethylene	Inexpensive, easy to shape	Not good for impact loading	Very easy to cut/machine	Heat staking

Figure 5.2 shows the first prototype of Measudrill. I combined two used steel brackets with aluminum rivets and then hammered the rivets for better joints to make the Measudrill body. Then I ground the excess aluminum rivets and polished the Measudrill body. I drilled a hole in the body and fitted a drill bushing with a 7/16-inch drilling hole. For the ruler marks on the body, I printed them on paper and then used duct tape to attach the paper to the body. For the guides that travel along the body, I cut aluminum sheet metal in several stacks and then bonded them. I drilled holes in the guides and used tap screws for clamping the guides on the body. When I tested the working prototype shown in figure 5.3, the results were impressive. The prototype was capable of drilling holes from any reference point of the material within 0.01 inch (0.254 millimeter).

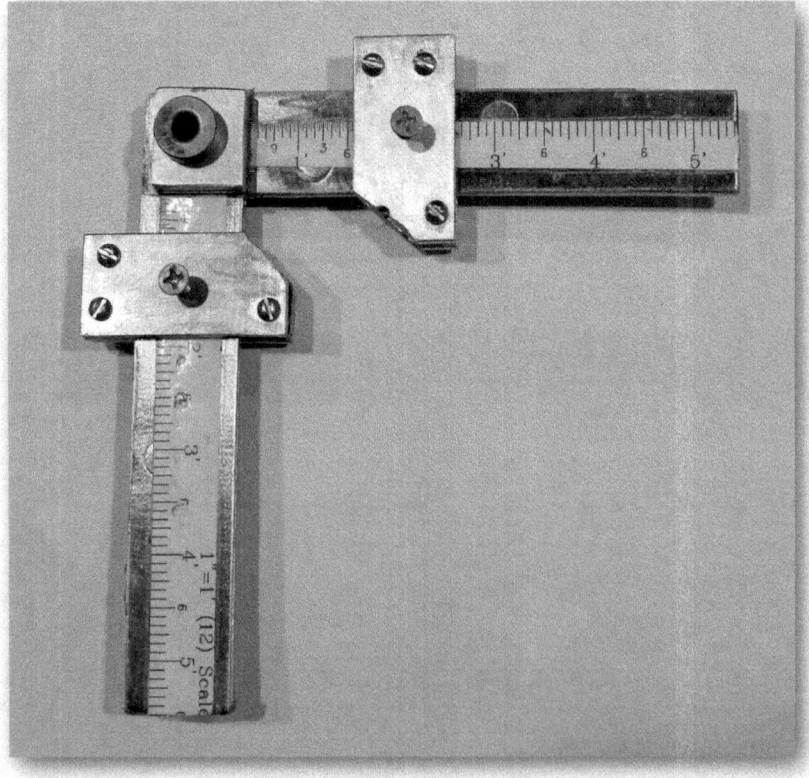

Figure 5.2. First prototype

Figure 5.3. Testing the first prototype

After creating a prototype, it is essential to get feedback from the target customer base. Feedback will help improve the design faster and reduce the design iterations from prototype to final product. You should set up demonstrations with the potential reviewers. Note their backgrounds and expectations. You can evaluate what they like about the product and use this information to refine the design and manufacturing of the product. I would suggest reviewing for three areas of positive feedback and three areas of negative feedback. If the negative feedback is dominant, I suggest taking the feedback seriously and reconsidering the design of the prototype. Once you complete the revised prototype, you can schedule another meeting with reviewers to ensure that the revised prototype meets their general expectations in terms of function, look, and usability.

Planning for prototyping in a large organization requires cross-functional discussion and consideration of the following additional areas:

technical feasibility, financial feasibility, suitability, technical evaluation and market research analysis, prototypes for technical and market testing, and product launch. Sometimes different departments in large organizations require different specifications for the prototype. It is important to meet the expectations of all departments by considering cost, function, and customer benefits.

Because most prototypes are not considered the final product, the cost calculations are often disregarded. However, a well-produced prototype is a good bridge to a more marketable product. It helps by reducing the number of iterations to the final product and simplifies the manufacturing process. During prototyping, it is best to choose the manufacturing process that will be used in the production of the final product—for example, manufacturing a lightweight ladder prototype with aluminum extrusions and then using the same aluminum extrusion profile and tools for the mass manufacturing. This practice can help reduce the cost of manufacturing by reusing the tools created in prototyping.

CHAPTER 6

COSTING FOR FULL-SCALE PRODUCTION

*If we could sell our experiences for what they
cost us, we'd all be millionaires.*
—PAULINE PHILLIPS

In This Chapter

- Cost calculation for manufacturing the product
- Determining profit margin for the product
- Designing a cost-effective product

Once you make a successful prototype, the next step can be even more complicated. Can you manufacture this prototype with an appealing design and packaging, at a high volume, at a reasonable cost, and at a profit? This chapter will address these issues. The first step in high-volume production is to come up with a sales price point for your product in the marketplace. In an environment where the customer's buying power is shrinking and there is high competition in the market, a proper product sales price target can eliminate costly design and manufacturing iterations. The flowchart in figure 6.1 shows the recommended steps for high-volume production planning.

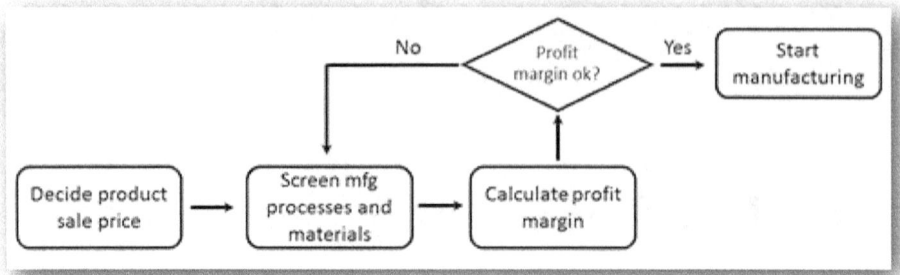

Figure 6.1. Flowchart for the manufacturing process

Product cost is the most essential element for a business because consumers want the lowest prices possible, and profits are determined by calculating price less costs. It is important to keep the cost down for all products, but it is essential to ensure that a new product has a competitive price to attract customers and build a customer base. Of course, if the price is too low, you could lose money, and the product might be not profitable or able to be sustained in the marketplace for long. There should be a balance in cost-effective price strategies, and these strategies are explained in the next section.

6.1. Pricing

Pricing is not easy task; it requires market study, competitor analysis, and customer analysis. Studies show that 5 percent more customers buy a product at $19.99 than at $20.00 (Raju & Zhang, 2010). When determining the pricing, it is important to focus on three Cs: customers, competitors, and company: the price should be affordable for the customer, competitive with other products on the market, and profitable for the company.

A careful pricing strategy can boost the maximum product value and sales volume. Without a pricing strategy, it is possible to lose market share against competitors or to create insufficient profit and put the business at risk.

One common pricing strategy is to take a lower profit and sell the product at a lower price with a much higher sales volume. In this case, you can dominate the market in the initial market condition. But if competitors follow the same strategy, it will be difficult for both your products and your competitors' to make a profit. An example is tablet technology. The first iPad was released by Apple in 2010 (Tablet Computer, 2016), and the average price of a tablet was about five hundred dollars. Since then, a number of other companies have released tablets, and you can now find tablets for as low as fifty dollars. With such a low price, it is almost impossible for companies to make a consistently high profit from tablet products. Another example is Dollar Store's products. I purchased several sets of personal care accessories, including scissors, a nail clipper, and a file, for one dollar each. Of course, they were made in China. I gave one of these sets to a friend visiting from India; he was surprised by the low price. Can the Dollar Store or the manufacturer make money from this product? I suspect that the profits are minimal, but the store has bought in bulk and hopes that you will buy multiples because of the low price or that you will buy other products that may have a better profit margin.

Determining the product cost and sales price is often a complex process for inventors, and it is necessary to have an understanding of basic pricing strategies. There are several ways to determine the cost and price of a product, but the three most common are cost-plus pricing, competition-based pricing, and consumer-based pricing (Raju & Zhang, 2010).

Cost-plus pricing is a simple process in which the price is determined by the cost of the product plus some margin, as the name implies. It is a fair process and financially prudent because it ensures profitable sales. In other words, it saves the company by taking care of profitability. One drawback of this pricing strategy is what happens when the product cost is further reduced. Also, if the company is able to produce a high profit with this strategy, there is no need to reduce the cost to boost sales and make the product more affordable for customers.

In **competition-based pricing**, you simply identify the competitor's price and set the price of your product at about the same level, plus or minus a few percentage points. The issue/risk is that this pricing lulls the price setter into passivity, so it doesn't control the price. Also, when double monitoring of the price happens, both companies' profitability can be at risk. If the company sets the price too low, not only the company but the whole industry can be jeopardized. Profit margin should not be reduced from 10 percent to 1 percent. For new products, there may not be competition, or the competitor's products may not have the same or similar features, so it's important to differentiate the product from the competition.

Consumer-based pricing is based on how much each customer is willing to bear. This is a favorite technique of car salespeople. They talk with the customer first and determine what he or she can afford. Then they show the customer the best car available in his or her budget. This gives them the flexibility to charge different prices to different customers, raising or lowering prices to match the size of the customer's wallet. The issue with consumer-based pricing is that the customer may ask for a low price and a higher value. For mass-market products, for which customization is often difficult, it is also difficult to apply this pricing strategy. Instead, you can manufacture products with different levels of features and define different prices. In this case, customers can buy the product according to their budgets.

When it comes to the right pricing, it is always important to ask the following questions:

- Cost: Can you further reduce the product cost?
- Price: How much can you change the price? How do you know the price is right?
- Profit: Does the price provide enough profit to the manufacturer, vendors, and distributors?
- Promotion: Can a promotion provide saving benefits to the customer as well as to the distributors by increasing sales and lowering costs? Think about the coupons that you receive every day.

Another pricing strategy is to offer a free gift with purchase. *Time* magazine often offers a free item as an incentive when you subscribe to the magazine for a year. Harbor Freight Tools, a discount tool store, also offers free items and discount coupons to entice you to come to the store and buy other items at discounted prices while getting the free item. Also, special sales are common, like Black Friday, where door-buster deals are created from the products and or product bundles designed by predicting massive sales volume and reduced sales price for customers.

Most distributors or retailers make little profit with this strategy, but the idea is that customers come to the store to buy these door-buster items, plus other products that are more profitable for the store.

For service- or training-related products, it is often difficult to know the need and buying power of customers, making it difficult to determine the right competitive price. Therefore, **pay-what-you-want pricing** may be effective. In this pricing strategy, the customer can decide on the price of the product. For example, Radiohead, the English alternative rock band, announced in 2007 that it would let fans decide how much they would pay for downloading a new album, *In Rainbows*. This strategy created a lot of attention from the media and customers at that time. Meanwhile, pay-what-you-want pricing can generate publicity, but it assumes that you have a product with low marginal cost, a fair-minded customer, and strong relationship between buyer and seller in a competitive market.

Some companies also offer products and services for free. It is often difficult to understand how these companies can make profits. Google offers a number of free products. The company started offering free e-mail and file storage services, and now it offers more than 150 free products (Mahen, 2014). How does Google make money? The answer is simple: more users mean more information from the users. Google collects anonymous data about you and uses this information for targeted advertisement services. In other words, the more information, the better-targeted advertisement they can provide. Google is then paid for the advertisement when you click on it and visit the advertising website.

Another important pricing strategy is **variable pricing**, in which the cost of the product is changed at certain times of the year. Holiday seasons are a good example. Most customers assume that retailers will have some incentives with products so that the sales volume will increase. Also, some retailers look for higher sales volume and products with a low cost or price. If all retailers follow this trend, it often leads to pricing wars, in which the product sales price becomes much lower than it should be. Price wars are a last resort, as they not only destroy competition but can also reduce the profitability of the product and possibly bankrupt the company. For example, there is a price war among the fast-food companies McDonald's, Burger King, and Wendy's (Peterson,

2015). When Wendy's started offering a "4 for $4" meal that included a cheeseburger, four chicken nuggets, small fries, and a drink for four dollars, McDonald's and Burger King responded with dollar-menu deals. These lower prices may be good for customers, but all the fast-food restaurants are expected to be hurt by this price war.

6.2. Other Pricing Strategies

Select holidays and major events are perfect opportunities for retailers whose customers buy products as gifts. These types of events are perfect for the major consumer-product manufacturers, who often have special products for these events. These are usually called **burst products**. For example, the day after Thanksgiving, some products sold are considered burst products. They are special products or product packages that are designed to sell more in volume and attract customers to the store to buy more products.

Another example is continually changing inventory and reducing prices. Some retailers—for example, Filene's Basement—use aggressive discount strategies (Raju & Zhang, 2010). The product price is reduced by 25 percent after twelve days, by 50 percent six days later, and finally by 75 percent compared to the list price. Six days later, the product is automatically given to a charity. Only 0.05 percent of inventory ever needs to be given to charity. Filene's sells five hundred thousand dresses a year, 90 percent of them before the first scheduled markdown.

Attracting more and specific types of customers to a single product can also be accomplished by opening auctions with an extremely low starting price. For these types of products, you need to have something unique, like a house in real estate or an antique coin on eBay. You can offer the product for the best offer, starting with a very low price. For example, in a recent auction in a real estate market in Canada, a house started at $250,000 with various interested buyers. The final price of the house was $325,000. When the buyer was asked whether he had made a good decision, he said, "No, but I win and that matters." He found a way to connect to the house. Could he have found another house with a cheaper price? Yes, but he wanted that house. Similar bidding wars happen on eBay. The most valuable items start at a low price to attract several bidders. In the end, the price may be much higher than the typical sales price.

Priceline often uses a name-your-price methodology. It sounds simple, but when you are not happy with the service, you are stuck with it. During my postdoctoral study, I rented a hotel room in Gainesville, Florida, for a week through Priceline. When I arrived, I found that the hotel room was not clean or maintained properly, and there were even bugs in the bed. When I asked the hotel management for a refund, they said the hotel was booked through Priceline, so they couldn't refund my money. I was stuck with the purchase! After getting a little frustrated, I found another hotel, but this time, I negotiated with the hotel management that if the hotel was not up to my standards, they would refund my money.

Subscribe and save is another strategy that is gaining popularity. It is also called pricing for marketing profitability. **Costco** uses this strategy; it charges a yearly membership fee to offer lower prices for products. **Amazon** uses Amazon Prime memberships to provide free shipping for Prime-eligible goods purchased, offer television shows and radio shows for free, and provide a food shopping service.

In some cases, consumer purchases are not based on pricing; there are other factors driving the purchase. For example, McDonald's does not have the best-tasting food, but they have created an environment that is appealing to kids, and people know what to expect at any McDonald's worldwide. Another example is in the wedding-related services industry. When planning their weddings, most people plan for a dream day and often buy rings, flowers, cakes, dresses, and so on all at huge markups, but consumers pay it because it is for their weddings.

Walmart uses opening price theory, also known as differentiated pricing. The company offers products with various price ranges that can fit everyone's budget. Whenever customers enter the store, the lowest-priced products are seen immediately, and seeing such low prices can convince customers that they can find a product within their budgets and encourage them to shop throughout the store.

There are many pricing theories out there. However, most pricing strategies become successful if they are backed by experience and instinct. Deep customer understanding, good economic intuition, and a healthy dose of street smarts are crucial for successful product sales. After determining the price target and potential profit of the product, the next step is to design the product for high-volume production.

6.3. Calculating Unit Cost

Before finalizing the design for high-volume production, it is a good idea to set a boundary for the cost of the unit. Unit cost can be determined in two ways: using contract manufacturers and asking them to quote the product cost or setting up a manufacturing environment to make the product on your own. The product cost for using contract manufacturers was shown as an example in section 4.5. The required investment to open a manufacturing facility was also discussed in that section.

If you would like to manufacture the product on your own, the manufacturing costs of the product have three categories: component costs, assembly costs, and overhead costs. **Component costs** may include standard or custom parts or both. Standard parts are typically manufactured by another company as a regular item and do not need significant customization to build your product. For example, a blow motor does not need to be manufactured as a special item for an air conditioning unit and can be purchased as a standard product. Custom parts require additional manufacturing processes to integrate it into the product assembly. For example, to manufacture a bracket, you can buy sheet metal and create a final shape after stamping or bending operations. **Assembly costs** are the costs of assembly operations. They differ from component costs but are added to the total cost of the product. **Overhead costs** occur during materials handling, quality assurance, purchasing, shipping, receiving, and equipment handling.

The easiest way to calculate the unit cost of a product is to add the costs of the components, assembly, packaging, and other expenses. Table 6.1 shows the unit-cost calculation for Measudrill and also shows the estimation of the low and high limits of the part costs. Estimation of hard costs such as rental fees, equipment, and power was already covered in table 4.3 of section 4.5 and can be reviewed as the additional costs of running the business.

Table 6.1. Estimation of product cost for Measudrill

Components	Quantity	Each (low)	Each (high)	Total $ (low)	Total $ (high)
Body	1	$0.80	$0.80	$0.80	$0.80
Guides	2	$1.00	$1.00	$2.00	$2.00
Bushing	1	$0.60	$0.60	$0.60	$0.60
Bushing housing	1	$0.20	$0.20	$0.20	$0.20
Drill bit	1	$0.55	$0.55	$0.55	$0.55
Assembly ($20/hr)		5 min.	10 min.	$0.65	$3.30
Packaging ($20/hr)		1 min.	5 min.	$0.33	$1.66
Total				$6.13	$9.11

6.4. Profit Margins

Making a profit is the target for any product being designed and sold in the commercial marketplace. Higher profit is better for the business, if the product can generate enough sales volume. Determining a profit target in the early stage of full-scale production is very important for a successful product because you must identify and choose suitable materials and manufacturing processes to achieve the proper product cost to yield the desired profit.

Some products provide a high margin for retailers, including jewelry, music, consumer electronics, health and beauty products, movie DVDs, toys, video games, men's and women's apparel, computer software and hardware, and books (Roy, 2015). It often puzzles the manufacturer what the profit margin for a product should be. Table 6.2 shows the profit margins for manufacturers, wholesalers, and retailers.

Table 6.2. Approximate profit margins (Ulrich & Eppinger, 2000)

Product categories	Manufacturers	Wholesalers /distributors	Retailers
Automobiles	10%–30%		5%–25%
Computers	15%–50%	8%–15%	
Appliances	20%–35%		
Consumer electronics	20%–40%		15%–35%
Sporting goods	20%–50%		
Industrial equipment	20%–45%	15%–35%	
Medical devices	35%–60%	15%–40%	
Toys	40%–60%	10%–20%	
Branded packaged goods	40%–70%	8%–20%	15%–35%
Consumer software	70%–90%		

As an example, the typical profit margins for industrial equipment based on table 6.2 should be the following:

- If you are selling the product directly to the customer with a 40 percent margin target, the cost of manufacturing should be no more than 60 percent of the expected sales price.
- If you are selling your product via a retailer that has an additional 45 percent margin target, the manufacturing costs should be no more than 33 percent of the expected sales price.
- If you are selling your product via a distributor who has an additional 20 percent margin target, the manufacturing cost of the product should be no more than 26.4 percent of the expected sales price.

In this case, an example product that is priced at twenty dollars to the customer should cost the manufacturer this much:

- Direct sales: $20 (1–0.40) = $12.00 cost to manufacture
- Retail sale: $20 (1–0.40) (1–0.45) = $6.60 cost to manufacture
- Distributor sale: $20 (1–0.40) (1–0.45) (1–0.20) = $5.28 cost to manufacture

These are the common profit margins and only a starting point for the pricing and cost of a product. For additional product details, it is important to adjust the price based on the market conditions and state of the economy. The longer the chain of distributors and retailers, the higher the cost and lower the profitability. For higher profitability, it is best to keep the supply chain between the customer and manufacturer as short as possible. After figuring out the profit targets, the next step is to choose materials and a manufacturing process.

6.5. Design for High-Volume Production

High-volume production steps require a careful assessment to keep the final product cost as low as possible. The manufacturing processes should be chosen based on the shortest amount of time between input and output and the least amount of waste. The design also needs to be checked for function, quality, ease of use, simplicity, clarity, order, aesthetics, innovation, and truthfulness of the final product. To further reduce the unit cost, you need to reduce assembly and packaging costs. Some manufacturing processes require manual operations, in which the parts need to be loaded into a certain machine to do the manufacturing and then unloaded and passed to the next manufacturing operation. Automating such manufacturing operations can decrease the time to manufacture products and reduce the costs of the parts and the end product.

If you decide to outsource manufacturing, you won't have significant negotiating power if you are an individual inventor. Large corporations have better negotiating power because they are much larger in the market and are more desirable because of their good reputation and higher-volume buying power. If you are a small business, these are the disadvantages when dealing with contracting companies. It is important to create a careful manufacturing plan to launch the product after cost reduction. Without cost reduction, the total profit from the product could be very low and jeopardize the business.

During the early product design for high-volume production, creating a specification is the first step. This specification should define the functional requirements and customer expectations of the product. The specification covers the essential requirements, including the following:

- **Functional requirements:** how the customer will use the product during the product life cycle. Various testing conditions in the user environment are selected to validate performance. Some of these conditions include corrosion-humidity testing, in which the product is tested in an aggressively corrosive and humid environment to ensure the durability of the product. If the product will be used in an aggressive handling environment, drop testing may be used. If the product will be exposed to chemicals, solvent resistance of the logo and marks should be tested under alcohol, grease, cream, or WD-40, depending on customer requirements.
- **Quality requirements:** the quality control steps to weed out defective products before they go to the customer. They also cover the inspection of the raw materials to make sure the incoming materials are in fact what was specified.

In production planning, choosing the right manufacturing process can reduce the product cost significantly. Stamping, cutting, and extrusion processes are a few processes commonly used for reduced cycle time. Figure 6.2 shows a progressive die-cutting process to cut a shape from a flat steel sheet and then punch a hole in it. The total cycle time could be less than a few seconds. Complicated stamped parts are made with this method due to fast cycle time. Other examples include bolts, screws, and nails. These are made by cold-heading process with an extremely fast cycle time (a couple hundred in a minute) and provide a very cost-effective product with no secondary operation needs.

Figure 6.2. Progressive die cutting. First the shape is cut from the sheet stock, and then a hole is drilled or punched.

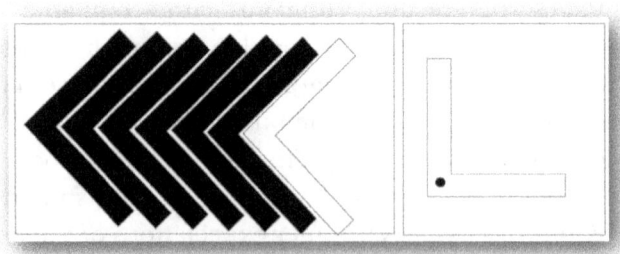

When the designer and manufacturing team select the manufacturing process for production, the candidate materials are screened for cost as well as manufacturing complexity. The total cost of the assembly is important, as a cheaper raw-material price may be cost effective, but complex-assembly costs can increase the final product cost. For products with complex shapes, casting is an economical manufacturing process because it reduces secondary machining operations and limits the number of parts that are required for assembly. Also, the cycle time for the casting process can be short, and complicated shapes can be integrated into a consolidated assembly. Most molds for metal casting are expensive, however, and may be too expensive for projects with small budgets. If the component can be made from various inexpensive steel grades, cutting and drilling operations can be economical to keep the cost down.

As a cost reduction example, one of the components in Measudrill, the bushing, was six dollars per unit on the market as a standard product. Instead of the bushing, I used a heat-treated, high-strength steel cylinder with the same inside and outside diameters. The cost was two dollars per foot. I cut it to quarter-inch lengths, and the cost was reduced to two dollars (12 x ¼ = $0.05). Compared to standard bushing, the high-strength steel cylinder was a much cheaper component alternative.

Materials used in assembly are also critical for the product cost and performance of the product. In most cases, steel grades are used in structural applications. One concern with steel is corrosion protection; steel needs to be painted or coated. An alternative material for structural applications is aluminum, which has become very popular due to its light weight and commercial availability. Because aluminum is lightweight, the component can be much lighter, if portability of the component is a concern. Table 6.3 shows the typical cost of various engineering materials. For the advantages and disadvantages of some of these materials, you can review table 5.1 in chapter 5.

Table 6.3. Cost of various engineering materials (Ulrich & Eppinger, 2000)

Material	$/kg	Material	$/kg
1020 steel	0.45–1.50	70/30 brass	2.10–3.00
1040 steel	0.45–1.50	#110 copper alloy	3.00–4.20
4140 steel	0.45–1.50	ABS	4.00–9.00
4340 steel	0.80–1.20	Polycarbonate (PC)	6.00–15.00
304 stainless steel	3.00–5.20	Nylon 6/6	6.00–11.00
316 stainless steel	4.00–8.00	Polypropylene (PP)	1.80–2.60
Gray cast iron	0.65–0.95	Polystyrene (PS)	0.80–1.40
2024 aluminum	3.80–6.00	Alumina ceramic	0.34–10.00
3003 or 5005 aluminum	1.20–5.50	Graphite	12.00–50.00
6061 aluminum	2.30–6.00	Douglas fir/pine	0.40–3.00
7075 aluminum	4.20–8.00	Oak	1.20–4.00
Magnesium AZ91D	1.20–2.80	Fiberglass/epoxy	2.00–11.00
Titanium 6-4	30.00–300.00	Graphite/epoxy	10.00–48.00

After choosing the manufacturing process for the components, the next step is to choose the joining process for the assembly. The joining equipment requirements and costs will add to the final costs of the assembly and product. Welding, riveting, and joining with bolts are commonly used joining processes. Riveting requires drilling a hole, removing the chips, and then a riveting operation to join the two parts. Another common joining process is welding due to its simplicity, cost effectiveness, and commercial availability. Steel and stainless components up to around 1.2 millimeters (1/8 inch) thick can be joined with spot welding without the need of filler material and shielding gas to eliminate the oxidation. For tungsten inert (TIG) or metal inert (MIG) arc, additional argon or helium gas is needed to create a protective environment for the welded metal.

While determining the joining process, the design can be simplified to reduce the number of components and the cost of making those parts and joining them together. Figure 6.3 shows the elimination of various parts and assembly steps achieved by design simplification. You can evaluate the design to see whether the mating components are of similar materials or have similar functions and if they are required for servicing or maintaining the product. If the mating components provide similar functions, it may be a good idea to integrate the parts for design simplification and cost reduction.

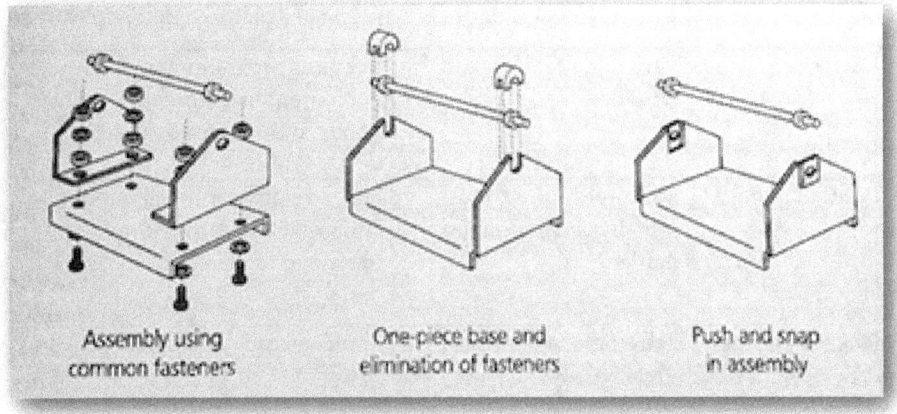

Figure 6.3. Design simplification example (Trott, 2008)

After identifying the right joining process, the next step is to select the right marking or labeling process for the product so that the company logo or additional marks can appear on the product. Table 6.4 compares the various marking processes. In terms of marking/engraving strategies for the logo or any other markings on the unit, CNC marking is a very easy process, but it can take ten minutes to make impressions on the part surface, so the cycle time can be longer than that of other marking processes. Pad printing is a simple marking process that creates markings with a rubber stamp. The cycle time can be two to five seconds because it uses a template and a rubber stamp to dispense the ink on the template. Screen printing is a marking process in which the parts are marked via a screen mesh template. It takes around five to ten seconds to mark the part. Ink-jet printing is widely used in marking manufacturing dates on boxes of sugar, cornflakes, and so on. It has an amazing marking speed of five meters per minute, and a part can be marked in less than two seconds. Laser marking has become a very popular marking technique. It uses a laser to etch the part surface, but it requires a painted or coated surface.

Table 6.4. Comparison of part marking processes

Process	Advantages	Disadvantages
Screen printing	Fast technique	Manual process
Pad printing	Fast and economical	Marking can wear off easily
Laser engraving	Very fine markings are possible	Requires programming; moderate equipment investment; aluminum needs to be coated for marking
CNC marking	Moderate speed; most metals, plastics can be marked	Requires programming; moderate equipment investment; marking tool need to be sharp
Ink-jet printing	Extremely fast process	Markings can rub off easily

Figure 6.4 shows the design iterations for Measudrill. In the first design, figure 6.4 (a), a slotted guide design was used. While this design had a great look, the locking function for the guides was not effective. In the second design iteration, shown in figure 6.4 (b), the guides were made as a closed system, and the clamping mechanism was achieved with a tightening nylon screw. The second design enhanced the durability and usability of the product during the drilling operations by giving extra stability to the guides and the Measudrill body.

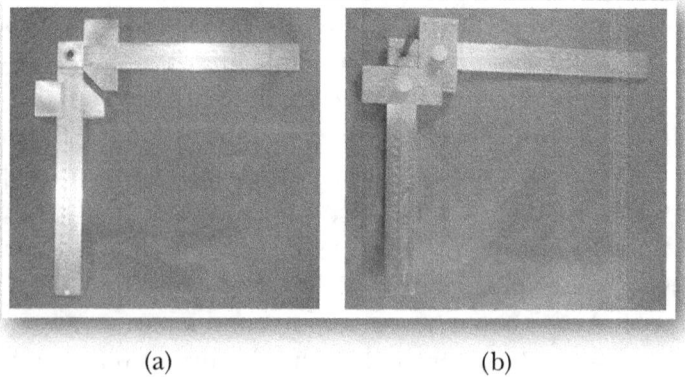

(a)   (b)

Figure 6.4. Measudrill design iterations. (a) First design; (b) second design

After determining the manufacturing processes for material, manufacturing, joining, and marking, next is logo and manual development.

6.6. Creating the Logo and Manual

A logo is defined as a symbol or other design adopted by an organization to identify its products, uniforms, vehicles, and so on. It is estimated that we see almost three hundred brand logos per hour in our everyday life (Douglas, 2011). A logo can also be used to communicate the product to the customer and can boost sales.

A logo should be simple and reveal the function of the product. The customer should be able to recognize the brand immediately and remember it. The colors of the logo should be simple and visually appealing. Fonts should be selected for readability and simplicity. Preliminary work is a must to create initial sketches. Balance between the size and amount of text and drawing is necessary. If the size of the text is too small, users cannot read what the logo is about. If the text size is too large, the logo can occupy too much space on the product and can distract the customer. The usage of color is important; do not use colors that are too bright or too dark. When the logo is printed in black and white, it should still look good. If a company is going to use the logo on its paperwork, it's also important to make sure that the colors used in the logo will copy and fax correctly. Enron's original "Crooked E" logo, along with all the stationery and business cards that used it, had to be scrapped when it was discovered after the fact that one of the colors the designer had used disappeared when a document was copied or faxed (Bierut, 2008). Design style should suit the product. If you have a logo, you should include the trademark information in the logo. Figure 6.5 shows the Measudrill logo.

Figure 6.5. Measudrill logo and sketch of how it works

You can either design the logo yourself or hire someone to create one for you. If you decide to do it on your own, you can use logo-design software. Most logo-design software costs as little as thirty dollars. However, most software requires a skill set and experience to create a visually appealing logo. If you do not have the experience or the patience to learn new software, you can hire a logo designer. There are many talented freelance graphic designers available at reasonable rates, and I recommend visiting Upwork.com. On this website, you post a project description, and then you interview freelancers to evaluate their previous design experience.

Before hiring a designer, you need to determine your budget as well as your expectations for the logo usage on the product and advertisement materials. Think about the type of logo you would like by considering your audience as well. Look at competitors' websites and at logos around you. Create a design wish list for the logo, and review this with the designer. Make sure that there is clear communication in the design iterations of the logo so that the final logo design is exactly what you want. Once you have the logo in the marketplace, it is a good idea to keep the design as is and not change it dramatically over time. This way, the customer can recognize the logo and product in the marketplace long term.

After logo development, the next step is to create a user manual to instruct the customer how to assemble and use the product. A user manual can be used to define the product functions, operating instructions, safety instructions, and warranty information for the customer. A well-developed user manual helps the commercial success of the product. When customers purchase the product, they should be able to quickly assemble it and understand how it works. They should know what kind of warranty they have if the product is defective. The manual should include the following sections:

- Name and a brief description of the product
- Contact information for the company
- Copyright information and notice for reproducing the manual
- Full description of the product
- Specification: detailed information about the product, dimensions, weight, materials used, and any other information about the product
- Assembly diagram and parts list
- Safety instructions
- Operating instructions
- Warranty information: what is and what is not covered

Some user manuals have only text and can be extremely frustrating for customers. Some manuals have only pictures, which can also be frustrating. For example, I assembled many toys when my daughters were born. I remember one particular toy, a climbing castle, that took me more than three hours to assemble, while according to the manual, the estimated assembly time was only forty-five minutes. I never bought another toy that required assembly from that company. There were no words, only poorly laid out pictures. Another example is furniture assembly. Most inexpensive furniture from stores like Ikea, Walmart, Target, and so on is produced in China and packaged unassembled to make the product as small as possible for transportation. When I was in college, I tried to assemble a bookshelf. It took an entire afternoon because the instructions were hard to follow and the pieces unlabeled. In other words, a poorly written manual can create a poor customer experience and may reduce sales. A good manual should have a balance between pictures and text and be clear.

The following checklist can help you create a successful manual and a good customer experience:

- Minimize the time required to read the user manual. Use bullet points rather than long sentences for the assembly and operating instructions.
- Use simple language. Avoid fancy words or technical terms.
- Focus on the customer who will read the manual, not on who is writing it.
- Include pictures or graphic representations in the assembly or operating instructions as needed.

Writing a user manual takes effort and constant editing to ensure the best customer experience. Adding pictures to the operating instructions can help the user quickly understand how the product works. Figure 6.6 shows the operating instructions for Measudrill in graphical descriptions. I took various pictures while operating the product and selected representative pictures for the manual. For a better demonstration of the product (and for marketing), you can also create a video and post it on social media sites such as YouTube or Facebook. This way, potential customers can quickly see what the product does and how to use it.

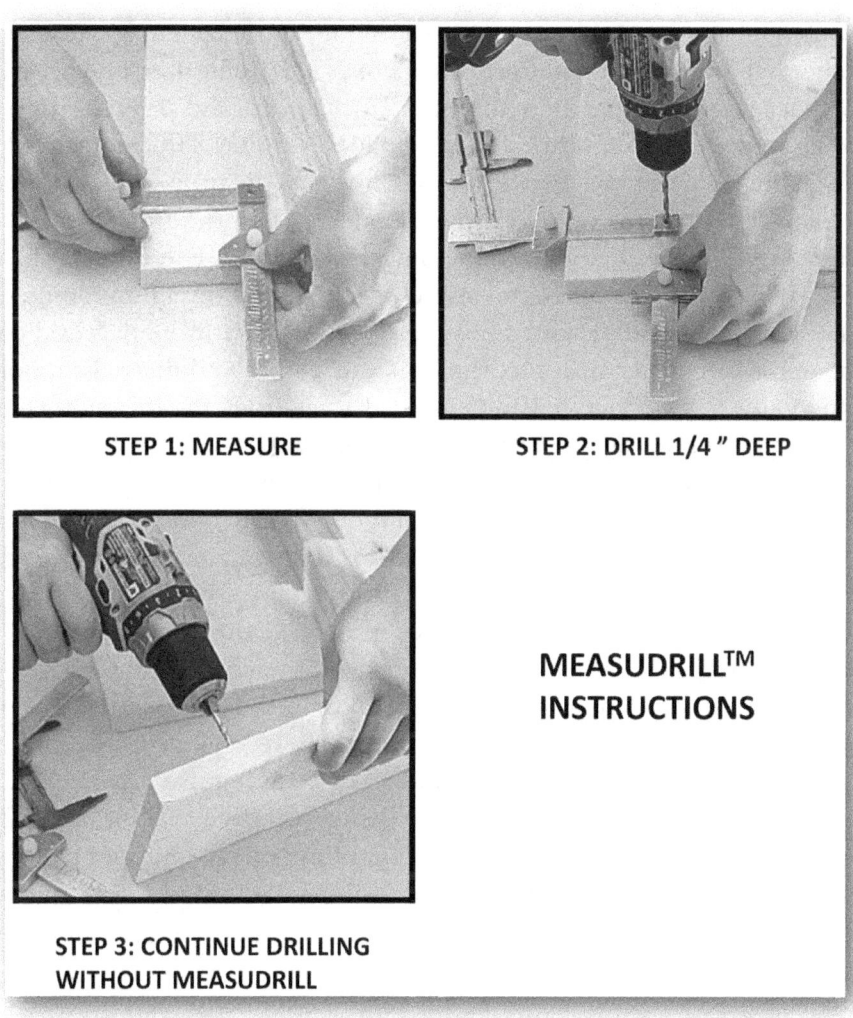

Figure 6.6. Operating instructions for Measudrill in graphics

6.7. Packaging

After creating the logo and manual, the next step is to choose the right packaging for the product. Packaging affects the marketability of the final product because both customers and retailers should like the packaging design. Packaging design should also protect the product against any damage during transportation and handling.

There are many choices for packaging, and the choice depends somewhat on the product. Cardboard boxes are widely used and are easy for packaging but may be expensive. In this case, the product dimensions are measured and fitted to the smallest volume. Thermoforming a packaging film may be an option to consider. This process uses a sheet of plastic film that is very soft when heated. While the film is soft, it is clamped to a fixture and vacuumed against a mesh. The film can form to the shape of the part between the plastic film and vacuuming mesh. The thermoformed plastic film is then combined with the part and corrugated cardboard to protect the unit and provide additional support. After gathering the manual and required accessories, the thermoformed plastic film is heat sealed for the final packaging.

Thermoform packaging has a very low cost compared to cardboard packaging, and investment for the equipment is very low. In fact, we built our own thermoforming machine for less than fifty dollars. Figure 6.7 shows the Measudrill in packaged form. During packaging, one design consideration was how to attach the product to the packaging so that it would look good after transportation. One way is to use tie wires to secure the product through predrilled holes in the cardboard. The cost of each tie wire is as low as eighteen cents. However, it requires drilling the cardboard. One solution I found was to use fugitive glue to attach the product to the cardboard, just like when you get a credit card in the mail. The cost of fugitive glue is as low as four cents, 4.5 times cheaper than tie wire and without the need to drill the cardboard.

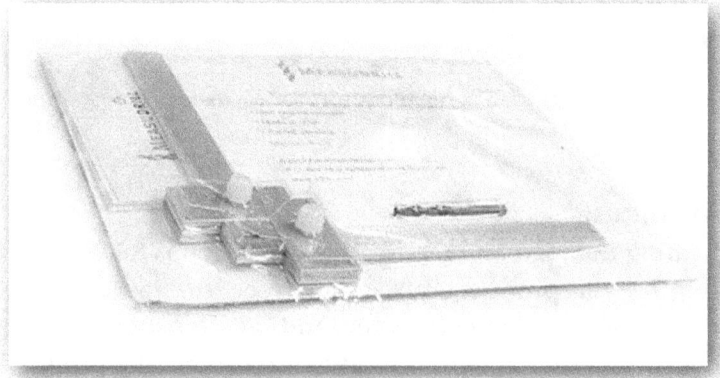

Figure 6.7. Measudrill in packaged condition

Other than thermoforming packaging, putting the Measudrill in a simple box and wrapping it with BubbleWrap was a possible packaging solution. Additional packaging techniques for various products are shown in table 6.5.

Table 6.5. Packaging systems (Kotler & Armstrong, 2006)

Packaging system	Product examples	Key attributes
Steel and aluminum tins and cans	Carbonated soft drinks	Not resealable (single serve); effective structure for graphics; recyclable
Folding cartons	Frozen cheesecakes, cereal boxes, Easter eggs	Versatile; final shape often a box but features such as handles can be added; cardboard engineering and new coatings provide additional opportunities
Hanging-pack formats	Popular within the DIY market	Inexpensive; ideal for small, low-cost items
Blister packs	Children's small toys, batteries	Versatile blister from PVC usually mounted on a backing card
Skin packs	Often used in promotions to put two products together (e.g., a jar of coffee and a packet of cookies)	Versatile blister from PVC, similar to above without the backing card mount; forms a "covering skin" around the product
Glass bottles and jars	Premium products, wine, baby food	Traditional; facilitates tamperproofing
PVC bottles and jars	Personal care products, carbonated drinks	More opaque than PET; less rigid; can have a handle incorporated; cheap; unbreakable
PET bottles and jars	Premium personal care products, carbonated drinks	Can be clear or colored; resealable; unbreakable; recyclable; rigid structure for graphics
Thermoform/fill/seal	Yogurt cups, pharmaceutical products	Simple; facilitates in-house packaging; cost effective
Composite containers	Pringles	Spirally wound paper-based tubes with plastic end caps
Bags	Potato chips, rice, sugar, fertilizer, retailer carrier bags	Wide variety of finished products available from high-quality paper carrier bag with rope handle to thin polyethylene carrier bag

Some packaging can be cost effective but also a nightmare for customers. More than sixty thousand people require hospital treatment each year from opening packaging (Britten, 2003). Also, new packaging materials can create issues for sales. SunChips tried 100 percent biodegradable bioplastic bags as chip containers. However, the bag was very noisy, and sales declined 11 percent. Due to declined sales and unhappy customers, SunChips bags were changed back to the original, quiet plastic bags (McDermott, 2010).

After reviewing the profitability of the product for high-volume production, create a schedule to ensure all the necessary steps for production are identified and captured. You can first define a work breakdown structure to outline the steps. The work breakdown structure as shown in figure 6.8 is a very useful tool to explain the steps to different production people and teams.

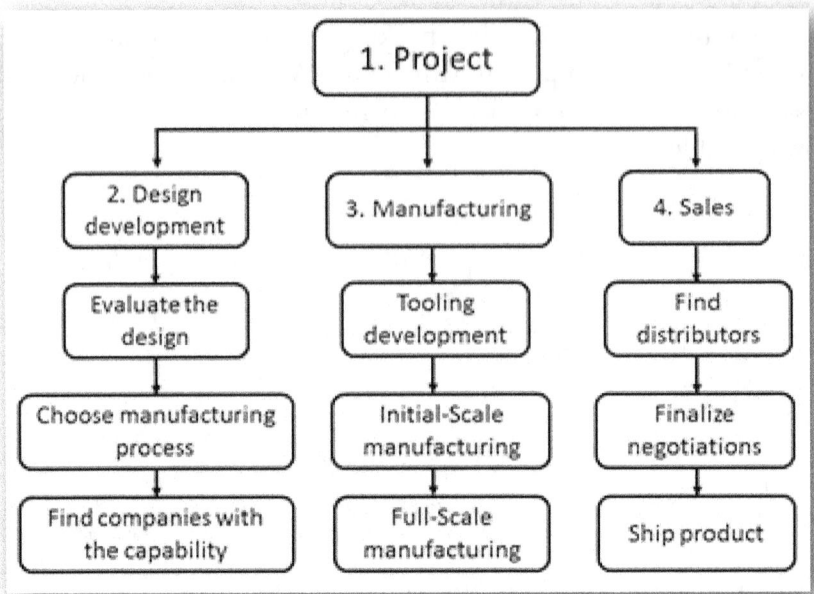

Figure 6.8. Project milestones for the next steps

As in prototype development, any project without a timeline may not be successful. A Gantt chart helps to outline the steps for final production. The schedule milestone for the production steps can be created as a Gantt chart, as shown in Table 4.4 in chapter 4. In this case, the critical tasks are to finalize the design, find companies for manufacturing the product, manufacturing starts, and first shipment of the product. While production is underway, the marketing plan needs to be developed to ensure that the product will be integrated into the commercial market easily. This step is discussed in the next chapter.

CHAPTER 7

MARKETING AND FINDING DISTRIBUTORS

Profit in business comes from repeat customers,
customers that boast about your project or
service, and that bring friends with them.
—W. EDWARDS DEMING

In This Chapter

- Creating a marketing plan for the product
- Creating a competitor analysis
- Finding distributors

Finding the right customers and selling the product at the right locations are as critical as manufacturing the product. Most inventors, unfortunately, make little effort to market their products. This is because marketing is an unknown area for them. Finding someone with marketing experience can also be difficult and often expensive. When I introduced Measudrill to the market, I tried to find marketing experts who could help me develop a marketing plan, but unfortunately, I could not find anyone in my initial search. Then I stopped looking for a marketing professional. I started with marketing books and checked real-life examples on the Internet. In fact, you can find many marketing books at reasonable prices to learn about the fundamentals of marketing. Some are listed in the recommended reading list.

It is essential to create a good marketing strategy to ensure the commercial success of the product. Without proper marketing, the product remains unknown and will be unsuccessful. Marketing has four key elements, the four *P*s: product, place, promotion, and price. Without a **product**, marketing is not possible, because the product is the starting point to develop a marketing strategy. As shown in the first chapter, the product can be a physical product, software, or a service. Products can be marketed in several categories; these are industrial and consumer goods, necessities and luxuries, durables and nondurables, convenience, shopping, specialty goods, and staple/fashion items. **Place** is the location where the product will meet the customer. Street vendors, supermarkets, and vending machines are all places. **Promotion** entices customers to buy the product. For example, running specials and discounts can help boost product sales. Finally, the **price** needs to be right in order to attract customers.

7.1. Marketing Plan

The marketing plan is the official document and road map for the product to be introduced and distributed to the market (Kotler & Armstrong, 2006). As mentioned, a good marketing plan should address the following elements:

- Type of customer the product is intended for and what values will be created for customer
- Product: what the product will be and what the primary functions of the product will be
- Price: how much the price for the product will be
- Place: what places and channels the product will be available from and how the product will be delivered to the intended target customers
- Promotion: how the offer will be communicated to target customers to persuade them of the product's merits

The marketing plan should also include a sales strategy. By combining the marketing elements, several marketing strategies are possible. These strategies can increase the sales volume of the product or improve the public image of the company. In the marketing plan, you can use the following marketing strategies:

- **Production concept:** the idea that customers will favor products that are available and affordable. For example, Coca-Cola is in vending machines everywhere. There are currently more than 6.9 million vending machines in the United States and 3.8 million vending machines in Japan (Shiozawa, 2015). Considering the population of the United States is 324 million now, there is a vending machine for every forty-seven people. Having more vending machines helps Coca-Cola's marketing strategy, as their products are available to customers almost everywhere.
- **Product concept:** the idea that customers will favor products that offer the most in quality, performance, and features and that the organization should therefore devote its energy to making continuous product improvements. For example, if you have enough money, you buy a BMW or an Audi because you have the perception that those brands provide a more comfortable ride.
- **Selling concept:** the idea that customers will not buy enough of the firm's products unless it undertakes a large-scale selling and promotion effort. For example, many men wait for Father's Day or Christmas specials to buy new clothes for themselves or the family because they know that the prices will be lower during those special holidays. Some people take sale specials to extreme ends. As shown in the TV program *Extreme Couponing*, there are people who spend their entire week collecting coupons, and when they go to the supermarket, they save almost 99 percent (Murphy, 2011). Is it good for them? If you need twenty cartons of pizza or thirty bottles of dish soap, yes, absolutely.
- **Marketing concept:** a concept for large companies with large budgets. The marketing management philosophy for achieving organizational goals depends on knowing the needs and wants

of target markets and delivering satisfaction better than competitors do. The concept starts with the sales planning, profits through sales volume by using integrated marketing, and aims for profits through customer satisfaction. For example, a large company can employ a market research team to gauge customer desires, use R&D to develop a product based on market research, and then use promotion techniques to ensure customers know the product exists. By using a variety of techniques, such as television commercials, billboards, and magazine advertisements, the company attracts more customers to boost sales.

Societal marketing: a new marketing concept popular with social media websites. It uses a principle of enlightened marketing that holds that a company should make good marketing decisions by considering customers' wants, the company's requirements, customers' long-term interests, and society's long-term interests. For example, using environmentally friendly packaging can significantly boost the image of the manufacturer.

With all marketing concepts, customer interaction is the key to success. You need to interact with the core customer types to identify their needs. When interacting with customers, it is important to record feedback. Audio recording, notes, video recording, and photography are effective tools to transfer findings to the marketing plan. Also, active listening is an important strategy for customer communications. Some customers may not directly reveal their preferences. You can ask starter questions about their previous experiences and the most frustrating or most liked products they used in the past.

After initiating the marketing plan, the next step is to do a competitor analysis to differentiate your product in the marketplace.

7.2. Competitor Analysis

A competitor analysis starts with benchmarking products of competitors. This requires a careful assessment of the functionality of the new product and a comparison with competitor products and their strengths and weaknesses. A marketing opportunity analysis, as shown in chapter 2, can provide an initial assessment regarding competitor conditions. To further evaluate competitors, the Thomas Register of America Manufacturers (www.thomasnet.com), Ariba Discovery (www.ariba.com), and Alibaba (www.Alibaba.com) provide directories of manufacturers of industrial products organized by product type. You can do a detailed search for competitive products to outline their objectives in marketing segments, product features, and cost structure. Benchmarking competitive products is an essential step for a marketing plan. Suggested resources are listed in the recommended reading list.

7.3. Finding Distributors

A good marketing plan also addresses the channels for distributing the finished product to customers. Having more distributors means you can sell the product to more customers. Here are some examples of distributors:

- Amazon sellers, eBay sellers, and trade shows. Most Amazon re-sellers are operated by between one and ten people. It is easi-er to talk to and work with them than with large corporations. Similarly, eBay has many sellers, and you can start selling the product yourself by opening a seller account. However, there is one drawback to eBay. The product will have limited visibility because it is new. Trade shows are another effective place to meet with distributors, both large and small. If you have the budget, you can sign up for a display space to show your product.
- After the product is launched, you can take it to the big box retailers: Target, Lowe's, Walmart, Menards, and so on. These stores are interested in highly profitable products with recognition. If your product is considered a specialized item, you may have little luck with these big box vendors.

There is a fundamental issue for finding distributors with new products. As described above, the big box stores in the United States usually don't stock a new product unless they know that there is interest from the public. Therefore, it is wise to start with resellers on Amazon or eBay. You can either ask for help from a seller or open up an account to sell the product yourself. You can ask resellers on Amazon and eBay to set up a sales space for your product. For artwork or creative works, you can use Etsy or Pinterest. The recommended reading list contains a complete list of websites for selling new products. For big box stores such as Walmart, Target, and Lowe's, you can check their websites to reach out to their sales teams and introduce your product to them.

If you are selling services, a marketing plan can also be useful. In this case, you need to focus on the type of service you are delivering and check whether any competitors deliver similar services. After creating a marketing plan for the product, the next steps should be to define an advertising campaign to inform potential customers and a sales plan to increase sales channels. These strategies are discussed in more detail in the next chapter.

CHAPTER 8

ADVERTISING AND SALES

In sales, it's not what you say; it's how
they perceive what you say.
—JEFFREY GITOMER

In This Chapter

- Advertising the product
- Creating a sales plan and strategy

No product can be successful unless the customer knows that it exists. Advertisement provides a communication platform to connect the customer and the product. Creating an advertising campaign requires careful customer assessment. There are challenges for identifying and reaching the right customer. In chapter 2 we identified the market segment, or the potential customer type, but now we have to get the product in front of those customers.

Advertising is the key element in reaching them. It is the promotion of the product and can be done by using the radio, the Internet, newspapers, or magazines. You need to create an attractive advertising campaign to increase the interest of the intended customer type. The marketing plan, covered in chapter 7, should help differentiate your product from the competition. The next step is to come up with a good, clear message to reach the customer through advertising.

8.1 Advertising

Advertisement is defined as any paid form of nonpersonal representation of goods, services, or ideas by an identified sponsor. Advertising usually does not use personal messages to promote a product. It presents informative and persuasive sales messages through the communications media, such as newspapers, television, and the Internet. It stimulates the interest in and demand for specific products, supporting sales promotion and personal selling. It is also used to generate a favorable public attitude toward a company, an industry, or another institution. For example, one of the best advertisement campaigns of all time was from Nike. With the "Just Do It" campaign, sales grew from $877 million in 1988 to $9.2 billion in 1998 (McCarron, 2007).

Advertising also helps new products by creating an initial demand. When a new product or service is introduced, there may be no demand because potential buyers do not even know the product is available. Advertising is the best way to inform potential customers about new products.

Before starting to advertise, you need to create an advertising campaign. The following categories are the primary areas of an advertising campaign:

- **By objective**: direct or indirect. Direct advertisement directly shows the product. Indirect advertisement establishes the brand and increases the acceptance of the brand. We often see seasonal commercials for companies during the holidays. This is considered indirect advertisement in which the company tries to increase the acceptance of the brand. Example: Santa with M&Ms or Google celebrating Chinese New Year. If you are just launching a product, direct advertisement is what you need.
- **By content**: product (product related) or institution (message of the company). For new products, product advertisement is essential because it introduces the product to the customer.
- **By market**: consumer or industrial. **Advertising to consumer** is the most important type but not the only type of advertising. **Trade advertising** is aimed at wholesalers and retailers. **Professional**

advertising is aimed at lawyers, doctors, and dentists, and **industrial advertising** promotes products for industrial use.

- **By geography**: nationally, regionally, or locally. Geographical advertising is often done in newspapers. If you require a larger audience, the Internet and TV are the best places to start. If your target is a local audience, radio, local newspapers, mail flyers, or Internet advertisements are the best.

- **By media**: print, electronic broadcast, direct mail, outdoor billboards, or the Internet. The Internet has become a major advertisement platform in this electronic age. It is possible to reach millions of people very quickly.

The objective and message of the advertising campaign should be clear to customers. They should support both personal selling, which means salespeople convincing customers to purchase the product, and indirect advertising to reach the customers whom salespeople cannot. Indirect advertising can open new territories for a company. When you advertise in a magazine or newspaper or on TV, the product may be viewed beyond your initial target audience. Using either direct or indirect marketing improves your relationships with vendors and consumers.

There are several ways of advertising your product and channels by which to do it (Bittel, Burke, & LaForge, 1984). **Word of mouth** is the first one. It is a very effective advertisement method; you just interact with the potential customers and let them know about your product. Friends and family members can be very effective and influential in most customers' purchasing decisions. The disadvantages are the limited number of people you can reach and the fact that some of them may not be the right customer targets.

The second method is advertising in a **newspaper**. You can reach a very large local market. The cost per thousand people reached is usually lower than other advertising methods, and you can revise the advertisement until the advertising deadline. The disadvantages are that customers read the newspaper quickly, and they skip some pages. It is also difficult to reach national audiences. Marketing segmentation is also not as precise with newspaper advertisement.

Magazines are another type of advertisement. They are suitable for reaching out to the national market, and they can provide very precise market segmentation. Customers also read magazines more carefully than they do newspapers, so magazine advertisements can be more successful than those in newspapers. The disadvantages include the longer lead time required to prepare an advertisement for magazine publication compared to a newspaper.

Television advertisement can provide a lively and persuasive national market, and it can be repeated frequently. The disadvantages are the high cost of advertisement for small advertisers. Also, advertisements must be prepared in advance and are difficult and expensive to change. **Radio** advertisement is similar to TV, but the cost is lower. The disadvantages are that the advertisement can be easily lost in program content.

Direct mailing can be used to directly solicit orders. Precise market segments of any size can be targeted locally, regionally, and nationally. However, cost per sale is high, and keeping the mailing list current may be difficult.

Outdoor advertisement can be used in high-traffic areas. Billboards repeat simple sales messages; however, they are not suitable for many products, and it is hard to keep the message up to date.

Internet advertisement can create high traffic from potential customers, and localized advertisement is possible. The disadvantages include the cost, which can be higher than that of other advertisement types. Also, most Internet advertisements are based on the pay-per-click (PPC) advertising model, in which the advertiser pays the publisher or owner of the website a certain amount for each time the ad is clicked. The effectiveness of the advertisement is measured by the number of clicks, as this shows the interest of the customer, although the clicks may not really represent targeted customers.

Another option is **trade shows.** Here you can present your products to the potential wholesalers, retailers, agents, or sometimes the general public.

You can also offer **samples** to the prospective customers or **small gift items** like key chains, pens, coasters, balloons, etc., to highlight and distribute your product and logo to potential customers.

The primary customers' behaviors and preferences define the advertisement media type you should consider. For example, a TV advertisement for a kids' yogurt product would be more successful than a newspaper advertisement because kids watch TV more than they read newspapers. Another example is lab equipment. An advertisement in a scientific magazine would be more successful than a TV advertisement because it would be much more difficult to reach lab equipment purchasers on TV than in a scientific magazine.

You can use a checklist to create effective advertising. Below are the recommendations for creating a successful advertisement:

- Be clear and use the customers' vocabulary.
- Provoke interest.
- Appeal to customers' wants and needs.
- Stress desirable features of product/service.
- Sound believable.
- Motivate people to buy.
- Tell them where they can buy.

The extent of the advertising campaign is completely up to you. The sky is the limit, but there are associated costs, so some careful budgeting is necessary. If you don't have any advertising expertise, you can work with advertising agencies. Advertising agencies are companies that plan, produce, and place advertisements in the media. You can search the Internet to find a local or global advertising agency. The Internet and online advertising are beginning to play dominant roles, and a list of places to advertise for free is in the recommended reading list.

I considered two types of advertisement for Measudrill. The first was a professional magazine advertisement. I determined that the target audience for Measudrill was woodworking professionals, and I tried to publish an advertisement in *Woodworking* magazine. Due to the high cost of advertising in a magazine, I also considered Internet advertisement. Google AdWords, Microsoft Bing Ads, Facebook, and Twitter are the common advertisement publishers on the Internet. For online advertising, I used Google AdWords to advertise Measudrill. You can choose different keywords for user searches and advertise your product in the search results.

I received 1.2 million impressions and 663 visits from the keyword "drill"; 544,000 impressions and 360 visitors from "hand tools"; and 98,000 impressions and 59 visitors from "wood working." I spent around $2,500 and received 2,200 visitors and 2.5 million impressions over a period of two months. If you want to run a larger advertising campaign, you can choose additional keywords for which you want your ad to appear in the search results. Figure 8.1 shows an advertisement for Measudrill.

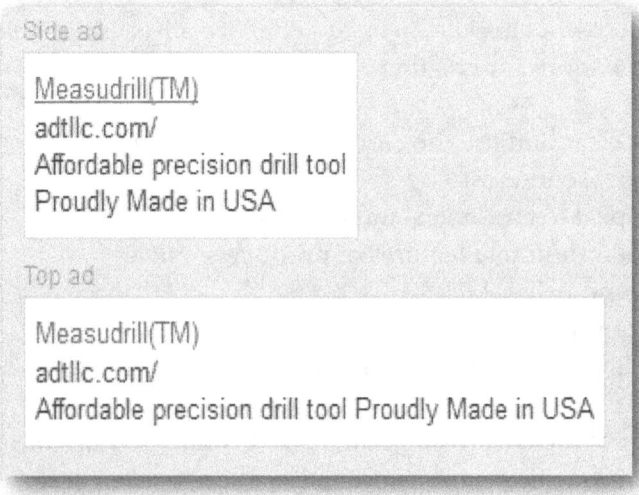

Figure 8.1 Advertisement example for Measudrill

The advertising material for the sales website should be very simple. On most e-tail websites, descriptions of the product are commonly written with bullet points for improved readability and simplicity. When you create a bullet list, it is recommended that the list is composed of parallel, consistent items. In other words, each is a sentence, each is a description, and the like. Below is the description for Measudrill:

- Precision drill measurement tool, patent pending
- Simple and practical design for accurate measurement before drilling
- Can be used with a wide range of drilling machines on metal, wood, or plastic materials

- Lightweight high-strength aluminum alloy frame, high-strength alloy steel drill bushing
- Includes drill bit (HSS steel with black oxide coating)
- Made in USA

Simple and descriptive graphical representations of the product are also helpful for the advertisement. Figure 8.2 shows an example.

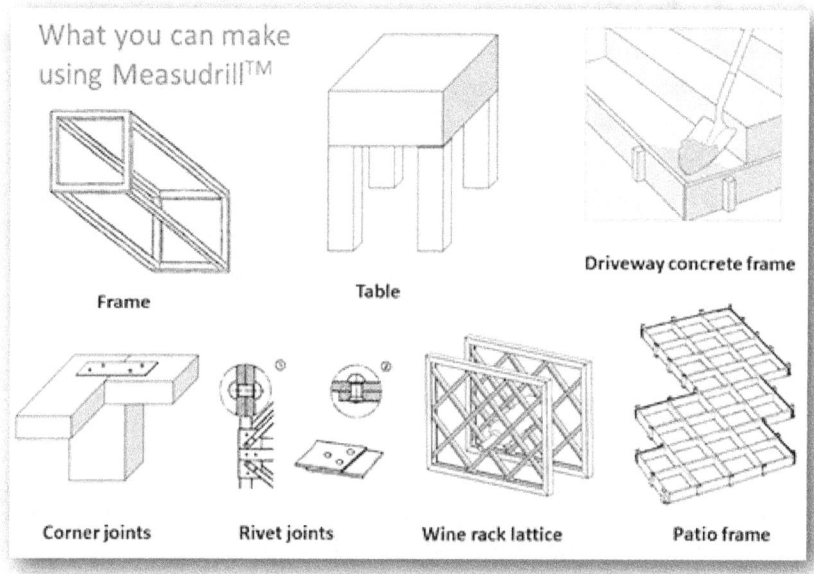

Figure 8.2. Simple descriptive advertisement for Measudrill

8.2. Sales

Sales are the lifeblood of business, so start thinking about the sale of the product as soon as you start designing it. Without sales, no business can survive. Every dollar spent on new-product development and production is against the company's/inventor's budget until the time that sales start. Figure 8.3 shows the typical projected cash flows during new-product development. Keeping product-development and manufacturing times as short as possible is a common way businesses can get the product to sales quickly and start making money.

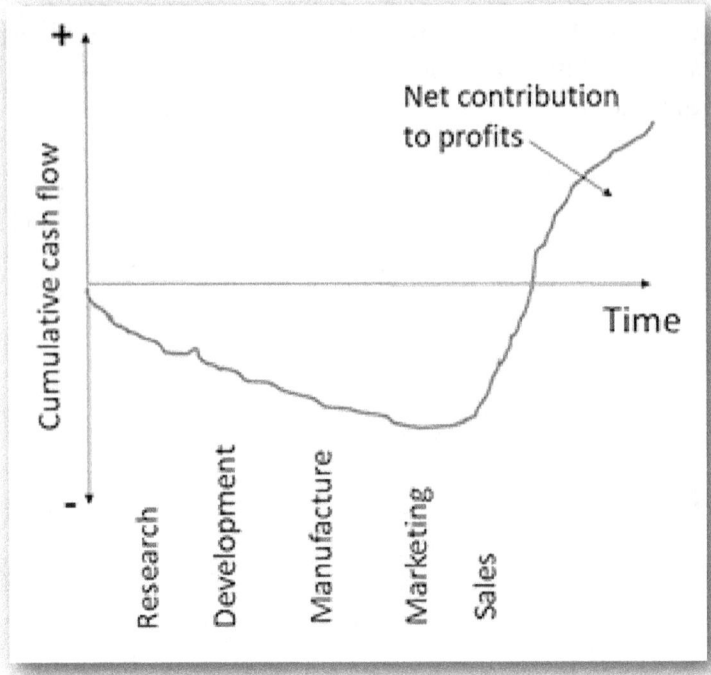

Figure 8.3. Cash flows in new-product development

To sell a product, you need to either be a salesperson or hire one, if you have the budget. Becoming a great salesperson is not easy. Can you sell a refrigerator to an Eskimo? It is a classic question about selling a product to a customer who doesn't need it. If yes, you are a great salesperson. Being a good salesperson requires confidence, pride, warmth, assurance, desire, enthusiasm, caring, and continuing education. You have to be comfortable meeting new people, doing research for potential customers, and making cold calls. It also requires the courage to face rejection. Motivators for successful salespeople are money, security, achievement, recognition, acceptance by others, and self-acceptance. For more information about personal development, refer to chapter 4.

Finding the right customers for sales is not easy, but it is not impossible. Sales can focus on individual/family, commercial, educational, and governmental sales. Reverse directories, club and company rosters, libraries, and mailing lists are good starts for individual sales.

Commercial and industrial sales resources to consider are Dun and Bradstreet's *Million Dollar Directory,* the *Thomas Register of American Manufacturers,* Standard and Poor's directory, and the *Who's Who* directories.

After creating the list of potential customers, you can further evaluate and prioritize the list by prospecting. Prospecting is a method of demonstrating the product to people you believe can and should buy it. You can reach potential customers via calling, e-mailing, filling out the "contact us" information on their website, reaching out to personal bloggers, mailing, or simply visiting them. Remember, every person on the list is a potential customer. You can use your business cards to reach people you have never met. Making more contacts leads to more potential customers. You should contact every potential buyer as though that person represented a thousand referrals. If you do, you will be selling your product by the hundreds or thousands in no time.

Preparation for customer interaction is the key element for sales. A short speech is believed to be the best and most efficient at all times. Research shows that after seventeen minutes, people lose interest, so be quick and to the point. Start by introducing yourself, the product, and key features of the product. Some openings are very common with salespeople, such as this example:

> Having useful gadgets in your garage and being able to invent something new are great feelings, aren't they? This gadget, Measudrill, is an exclusive product from our company to help you drill holes at accurate distances from x and y reference points without the need of expensive machining equipment. And it is built locally with environmentally friendly materials to provide the best value for our customers. I would like to show you the product today. It will greatly enhance your creativity when you own it.

When discussing the price, everyone likes a deal or special price. I recommend that you use an expression similar to one of the examples below, so customers feel that the product price is adjusted for their budgets.

It is available for $17.50.
It is valued at $17.50.
We offer it at $17.50.

Continuous improvement of the product is the key to long-term success in sales performance. When customers purchase the product, you should follow up to find out what they like as well as what they dislike about it. If they are not happy with it, ask them what they would like to see altered or improved in the new product. Some customers also like to be treated specially. Ask whether they would purchase the product again if the product could be customized to their needs, and provide options.

When it comes to rejection, we all take it too personally. However, rejection should be used as a learning experience, instructions/feedback for reaching goals, an opportunity to develop a sense of humor, or an opportunity to practice your techniques. Rejection is part of the game that we must play. Some salespeople believe that rejections are one more step toward being more successful.

Minor objections from customers require a follow-up if possible. If the customer doesn't provide enough information about what he or she dislikes about the product, you may need to ask additional questions.

Sales is hard work. You need to be persistent to find new ways to sell the product to the customer and find new distributors. If you receive a few rejections, it is OK and should be expected. Barbie dolls, photocopiers, vacuum cleaners, Monopoly, and Tupperware all were initially rejected in the market but are now very successful and part of everyday life.

There may be several reasons for rejection. First, the customer or distributor may not have understood what the product was or what it would do. In that case, you need to work on advertisement materials to ensure that your message is clear. Second, the sales price might have been incorrect for the market; it could be that the customer did not think the price was right or that the distributor or retailer did not think there would be enough profit on the product. In that case, cost reduction is essential. A positive outlook is essential for a successful new product. For additional information regarding sales strategies, please refer to the recommended reading list.

Remember, Thomas Edison made a thousand unsuccessful attempts at inventing the lightbulb. When a reporter asked, "How did it feel to fail a thousand times?" Edison replied, "I didn't fail a thousand times. The lightbulb was an invention with a thousand steps." I am sure you won't have to try your product idea a thousand times. With a little courage, discipline, and luck, you too can pursue your dream of creating something for society.

CHAPTER 9

CONCLUSION

*Focus on the journey, not the destination. Joy is found
not in finishing an activity but in doing it.*
—GREG ANDERSON

Congratulations if you have read this book up to here. I hope that you can use this book for future idea generation and product-development projects. Creating something new can be a truly amazing experience in which you can learn about your boundaries and push them further.

Commercialization strategies for inventions change based on product complexity and whether you are an individual inventor or working at a large corporation. For the most efficient product-development strategies, I have the following recommendations for inventors with limited prototype and manufacturing capability: Start with small-scale projects. Tools and home-market products are the easiest to manufacture. In some industries, like medicine, agriculture, and aerospace, it can take years to develop and test products. You can use the recommendations from chapter 1 to focus on the right industry for your new product idea. Also, many inventors and start-up companies need to develop shared services and manufacturing assistance to reduce product and operation costs. You can use such shared services or work with experts. If you don't have the budget to hire them, you can offer them sweat equity: some portion of the profit from the new product in return for their work. That way, you don't spend as much money during the development of the product and can allocate more money to advertisements.

For larger companies or complex product-development programs, my recommendation is to focus on invention rather than just innovation. Because large companies focus on their core businesses, they tend to channel their resources into the innovation or improvement of current products. Doing this long term can cause the company to miss opportunities to come up with new products. Constant engagement with customers can also help you understand the big issues that they are facing and maybe the new products that they are willing to pay for.

I hope that this book has also provided valuable tips to improve your personal skills, better prepare the product, find resources, and develop a team. Remember, nothing can be done on your own; you need to connect with others to ensure the success of the product. You need to have an idea, a team to work with, money to support product development, and discipline and determination to pursue your dream to create a commercial product.

In terms of Measudrill, I am quite happy with the results. I wanted to create my own product and experience all the transformation steps from the idea to the store shelf. After some steep learning curves, I finally sold a number of units and have received many positive comments about the product. Of course, I have listened to the negative and neutral comments from customers as well and used this information to improve the product. After the product was commercialized, I helped several local inventors make sure that they would benefit from my experience. Inventing and developing a product is all about the journey, what we can learn from the product and customers, and how we can build a product with the highest customer satisfaction.

ACKNOWLEDGMENTS

This book has been created with the help of many collaborators. I would like to extend my thanks to my wife, Melissa Ann Varlioglu, for her endless encouragement to finish my invention as a commercial product; James Anderson, who helped with making the prototype and with manufacturing; Stanislava Getova for creating an exceptional logo and artwork for my invention; Ali Kamil Varlioglu, who provided feedback about the manufacturing business; Ahmet Alpay, who was my first retail distributor; and my daughters, Ella and Leyla, for having a positive outlook at all times.

APPENDIX 1

GATHERING INFORMATION

For gathering information, you need to create customer interviews and read reviews of products already on the market. You should also read about new trends in the consumer market (iPhone, smartphone, cloud technology, etc.) and review the product contents and customer reviews in the product or retailer's website to learn about customer expectations. To sell a product, you need to identify what the customers' expectations are with the current product and solve unmet needs, even if that means selling less or recommending another company's products. By applying this approach, you can also identify the most common issues that customers are facing.

Information Gathering Template   Created By: [_____]

Date: [_____]

Product: [_____]

Detailed product information: [_____]

1. Customer Information:

Age Group: ( ) 0-4 ( ) 5-12 ( ) 13-18 ( ) 19-25 ( ) 26-34 ( ) 35-50 ( ) 51-65 ( ) 66-70 ( ) 71+

Gender: ( ) Male ( ) Female   **Loyalty:** ( ) First-Time Customer ( ) Returned Customer

Location: ( ) North America ( ) South America ( ) Asia ( ) Europe ( ) Africa

Product Type: ( ) Needs (groceries, diapers, essential items, etc.) ( ) Wants (toys, gadgets, etc.)
( ) Desire (Expensive cars, gold watch, etc.) ( ) Other: _____

Product Usage: ( ) Personal Use ( ) Industrial/Professional Use ( ) Training ( ) Gift
( ) Donation ( ) Other: _____

Purchase Method: ( ) Online ( ) Mailing ( ) Offline ( ) Other: _____

Purchase Time: ( ) Holiday ( ) Seasonal Item ( ) Personal Celebration (birthday, etc.)
( ) As a donation ( ) Other: _____

2. Purchase Evaluation: Check if the product fits with customer's needs

Price: ( ) Yes ( ) No ( ) Maybe ( ) Other: _____

Function: ( ) Good ( ) Satisfactory ( ) Fair ( ) Low ( ) Other: _____

Quality: ( ) Good ( ) Satisfactory ( ) Fair ( ) Low ( ) Other: _____

Expectation: ( ) Good ( ) Satisfactory ( ) Fair ( ) Low ( ) Other: _____

Buying Experience: ( ) Good ( ) Satisfactory ( ) Fair ( ) Low ( ) Other: _____

3. Usage Evaluation: Check if the product fits with customer's intended usage

Quality: ( ) Yes ( ) No ( ) Maybe ( ) Other: _____

Function: ( ) Good ( ) Satisfactory ( ) Fair ( ) Low ( ) Other: _____

Usage Experience: ( ) Good ( ) Satisfactory ( ) Fair ( ) Low ( ) Other: _____

4. Any recommendation for ideal product:

[_____]

5. Three improvement areas (function, price, look/feel, weight, and service performance)

Three best features: [_____]

Three improvement areas: [_____]

APPENDIX 2

CREATING PRODUCT IDEAS

A product idea can be created by starting with an existing product on the market. Defining the basic functions of the product is the first step. The next step is to define three weak areas and how you can improve the product in terms of function, price, look/feel, weight, and quality. You can create new ideas with this template and then pick the top five product ideas.

Idea Creation Template Created By: [   ]

Date: [   ]

Product Name: [   ]
Product Manufacturer: [   ]
Basic Functions of the Product: [   ]
Product Extra Functions: [   ]

Weakness Areas. Note 3 frustrating or disappointing areas from the customer.

[   ]

Idea Generation. Check for opportunities for product improvement or new product ideas. How can we improve the design?

Function: [   ]
Price: [   ]
Look/Feel: [   ]
Weight: [   ]
Quality: [   ]

Come up with at least three ideas and determine how attractive those ideas would be to the customer in terms of function, price, look/feel, weight, and service performance.

[   ]

Report the information in a presentable format and analyze if the information can lead to a new idea, product, or solution.

Idea 1:
Idea 2:
Idea 3:
Idea 4:
Idea 5:

APPENDIX 3

SCORING PRODUCT IDEAS

As discussed in chapter 1, you can score your ideas in terms of technical, competitive, patentability, and financial standpoints. You should use at least three ideas created in appendix 2 and further evaluate the best one. In the idea scoring section, score the ideas in terms of expected design efforts, cost/man-labor needs, technical difficulty, manufacturing difficulty, competition, and market attractiveness as high/medium/low grades. The overall judgments are calculated by the weighted average technique, and the ideas are ranked. If no idea gets a good score, there is no need to further evaluate them. You should go back and keep producing different ideas. For the best idea, you can evaluate the technical needs, competitors, patentability, market, finance, and future plan areas. For the critical questions with negative answers, you need to improve the conditions before making the final decisions.

Idea Scoring Template

Created By: ____

Date: ____

Function: ____
Target Industry: ____
Target Price: ____
Main Competitors: ____
Differentiating Factors/Features from Competitors: ____

Idea 1: ____
Idea 2: ____
Idea 3: ____
Idea 4: ____
Idea 5: ____

Idea Scoring

	Ideas	Expected Design Efforts	Cost / Man-Labor	Technical Difficulty	Manufacturing Difficulty	Competition	Market Attractivity	Overall Judgments	Rank
1									
2									
3									
4									
5									

Resource Assesment

	Technical	No	Maybe / Unknown	Yes
1	Do we have experience with the technology?			
2	Do we have skills and facilities to develop this product idea?			
3	Does it look like a probability for technical success?			
	Competitors			
4	Does this idea provide anything new compared to the competition?			
5	Is it necessary to defend an existing business?			
6	Is the product likely to be superior to the competition?			
	Patentability			
7	Can we get patent protection?			
8	Can we get a trademark for the product idea?			
	Market Analysis			
9	Is this a growing or stable market?			
10	Is there an existing customer base?			
11	Is the size of market reasonably large?			
	Finance and Future Plan			
12	Is the return on investment reasonable?			
13	Does it support our short-term and long-term plans for the business?			

Results: **Go** **No Go**

APPENDIX 4

MARKETING OPPORTUNITY TEMPLATE

The marketing opportunity template evaluates the product idea against the marketing conditions and competitors. You can use this template to check your idea.

Marketing Opportunity Template

Created By: [   ]

Date: [   ]

Product Idea: [   ]

1. Opportunity Story

Target Segment: [   ]

Value Proposition: [   ]

Customer Benefits: [   ]

Critical Resources: [   ]

Reasons to Believe: [   ]

Resource Sourcing: [   ]

How to Monetize: [   ]

Opportunity: [   ]

2. Attractiveness Factors:

Unconstrained Opportunity: [   ]

Segment Interaction: [   ]

Growth Rate: [   ]

Market Size: [   ]

Profitability: [   ]

3. Opportunity's Attractiveness:

	High	Medium	Low
Competitive Vulnerability:			
Magnitude of Unmet Needs:			
Interaction Between Segments:			
Likely Role of Growth:			
Technology Vulnerability:			
Market Size:			
Level of Profitability:			

APPENDIX 5

PERSONAL STRENGTHS

*T*he Basic Principles, developed by Zenger Miller, is a great resource to identify your personal strengths and weaknesses before starting a company. The survey below is a summary from his book and helps you to evaluate your personal strengths.

	Focus on the situation, not on the person			
	I make a conscious effort to...	Almost never	Sometimes	Almost always
1	Step back and look at the big picture when analyzing a situation.	1	3	5
2	Avoid letting personality differences keep me from dealing with a problem.	1	3	5
3	Make decisions based on facts.	1	3	5
4	Consider other points of view.	1	3	5
	Maintain the self-confidence and self-esteem of others			
5	Openly express confidence in others.	1	3	5
6	Recognize accomplishments and ideas.	1	3	5
7	Encourage people to express their ideas.	1	3	5
8	Encourage people to use and expand their abilities.	1	3	5
	Maintain constructive relationships			
9	Use every interaction as an opportunity to build relationships.	1	3	5
10	Acknowledge problems openly, honestly, and objectively.	1	3	5
11	Deal with conflicts as they arise.	1	3	5
12	Share information.	1	3	5
	Take initiative to make things better			
13	Look for opportunities for improvement.	1	3	5
14	Stay informed.	1	3	5
15	Act as if there is a creative a solution to every problem.	1	3	5
16	Ask for and offer help.	1	3	5
	Lead by example			
17	Model the behaviors I expect others to practice.	1	3	5
18	Follow through on my commitments.	1	3	5
19	Admit my mistakes.	1	3	5
20	Challenge myself and others to try new ways of doing things.	1	3	5
	Think beyond the moment			
21	Uphold ethical standards.	1	3	5
22	Weigh the impact of my decisions before taking action on them.	1	3	5
23	Set objectives that instigate action.	1	3	5
24	Plan ahead.	1	3	5

Check your responses in each section and determine your areas of strength. A score of 18 or more in a section shows an area of strength. If your score is less than 9 in any specific area, improvement is needed. You can get feedback from your coworkers, family, and friends after working on the area with the low score. Once you see improvement, I am sure they will recognize the changes and let you know right away.

APPENDIX 6

CHECKLIST FOR FUNDING EVENTS

B efore meeting with investors, it is essential to go over the items below and evaluate the strengths and weaknesses of your proposal. Review any weaknesses, and determine how they could be improved (VentureNet Evaluation Form, 2016).

Question	No	Maybe/ unknown	Yes
Business concept Market segmentation identified? Is there a market? Fulfills the marketplace need? Market feasibility studies indicate acceptance? Business concept is mature?	1	3	5
Business model Developed/articulated model? Pricing structure/rationale? Detailed historical financials provided? Projected cash flow strategy? Business model is mature?	1	3	5
Marketing strategy Differentiates the idea from competition? Defined marketing plan for product launch completed? Sales/distribution plan completed (including value chain)? Business development leadership in place or being recruited? Marketing strategy is mature?	1	3	5
Management team Founders have a realistic view of role in company? Leadership shows previous similar direct experience? Complete team with relevant knowledge and full-time commitment? Mentors and advisors (strategic, operational, legal, certified public accountant)? Management team has a mature structure?	1	3	5
Outside validation Assessment by industry expert? Market introduction/customers? Outside validation section is mature?	1	3	5
Technology/product/service Intellectual property opportunity/requirements studied? Property—intellectual property, trade secret, defensible? Unique and/or competitive? Completed/market ready?	1	3	5
Capitalization/funding program Proposed use of funds/milestones? Company taking on proportionate risk with need for funding? Financial plan in place with nonpublic funding match secured? Finance structure that aligns risks/incentives of investors with state funding? Capitalization strategy for follow-on private funding? Detailed capitalization chart provided? Capitalization/funding plan is mature?	1	3	5
Exit opportunity/strategy Anticipated defined goal? Timing/comparable/return (defined)? Exit opportunity/strategy rate?	1	3	5

REFERENCES

Anastasio, A. (2016, May 12). Dream funding. Retrieved from http://www.alexiaanastasio.com/work-with-me/

Atari video game burial. (n.d.). Retrieved May 5, 2016 from Wikipedia: https://en.wikipedia.org/wiki/Atari_video_game_burial

Barber, G., Birnbaum, R., & Lieber, M. (Producers) & Abraham, M, (Director). (2008) *Flash of genius* [Motion picture]. Canada: Universal Pictures

Berger, J. (2016, 7 26). Shopkins secrets and fun facts. Retrieved from Newsday.com: http://www.newsday.com/lifestyle/family/shopkins-secrets-and-fun-facts-1.12058684

Berkowitz, E., Kerin, R., Hartley, S., & Rudelius, W. (1999). *Marketing*. Homewood, IL: McGraw-Hill.

Bierut, M. (2008, 2 11). The smartest logo in the room. Retrieved from DesignObserver.com: http://designobserver.com/feature/the-smartest-logo-in-the-room/6237/

Bittel, R. L., Burke, R. S., & LaForge, R. L. (1984). *Business in action: An introduction to business*. New York, NY: McGraw-Hill.

Boothroyd, G., Dewhurst, P., & Boothroyd, D. (1994). *Product design for manufacture and assembly*. New York, NY: Marcell Dekker.

Britten, N. (2003, February 11). 60,000 are injured by opening packaging. *The Telegraph*: Retrieved from http://www.telegraph.co.uk/news/uknews/1421698/60000-are-injured-by-opening-packaging.html

Broom, H., Longenecker, J. G., & Moore, C. W. (1983). *Small business management.* Cincinnati, OH: South-Western Publishing Co.

Covello, J. A., & Hazelgren, B. J. (1994). *The complete book of business plans.* Naperville, IL: Sourcebooks.

Dlugan, A. (2010, August 22). *17 easy ways to be a more persuasive speaker.* Retrieved from http://sixminutes.dlugan.com/ logos-examples-speaking/

Dornin, R. (2007, May 23). 2 sentenced in Coke trade secret case. Retrieved from http://money.cnn.com/2007/05/23/news/ newsmakers/coke/

Douglas, S. (2011). The guide to great logos. Retrieved from http://www. thelogofactory.com/logo_blog/wp-content/uploads/2011/09/ Guide-to-great-logos-v1.pdf

Ferrell O. C., & Hirt G. (2000). *Business: A changing world.* Boston, MA: McGraw-Hill.

Fitzpatrick, D. (2015). Avoid trademark infringement when you choose a domain name. Retrieved from http://www.nolo.com/legal-encyclopedia/avoid-trademark-infringement-domain-name-29032.html

Gallo, C. (2013, May 16). How Warren Buffett and Joel Osteen conquered their terrifying fear of public speaking. Forbes. http://www. forbes.com/sites/carminegallo/2013/05/16/how-warren-buffett-and-joel-osteen-conquered-their-terrifying-fear-of-public-speaking/

Goodin, D. (2015, October 23). "Joomla bug puts millions of websites at risk of remote takeover hacks. Retrieved from http://arstechnica. com/security/2015/10/joomla-bug-puts-millions-of-websites-at-risk-of-remote-takeover-hacks/

Govindarajan, V., & Trimble, C. (2010). *The other side of innovation: Solving the execution challenge.* Brighton, MA: Harvard Business Review Press.

Govindarajan, V., & Trimble, C. (2012). *Reverse innovation: Create far from home, win everywhere.* Brighton, MA: Harvard Business Review Press.

Grow, K. (2015, March 10). Robin Thicke, Pharrell lose multi-million dollar "Blurred Lines" lawsuit. *Rolling Stone.* http://www.rollingstone. com/music/news/robin-thicke-and-pharrell-lose-blurred-lines-lawsuit-20150310

Hanke, M. (2013, October 14). 10 good techs turned bad. Retrieved from http://www.seeker.com/10-good-techs-turned-bad-1767944010.html

Hardigree, M. (2009, August 25). How Jalopnik reunited Papa John with his Camaro. Retrieved from http://jalopnik.com/5344739/how-jalopnik-reunited-papa-john-with-his-camaro

Hopkins, T. (1982). *How to master the art of selling.* Scottsdale, AZ: Champion Press.

Huen, E. (2016, April 25). Inside the Hermès Birkin bag that sold for record $298,000. *Forbes.* http://www.forbes.com/sites/eustacia-huen/2016/04/25/inside-the-hermes-birkin-bag-that-sold-for-record-298000/#4d07763f2dbb7

Janov, J. (1994). *The inventive organization: Hope and daring at work.* San Francisco, CA: Jossey-Bass.

Keppler, N. (2016, January 31). January 31, 1990: McDonald's opens in the Soviet Union. Retrieved from http://mentalfloss.com/article/74609/january-31-1990-mcdonalds-opens-soviet-union

Kotler, P., & Armstrong, G. (2006). *Principles of marketing.* Upper Saddle River, NJ: Pearson Prentice Hall.

Krassenstein, B. (2015, July 18). What is 3D printing & how do 3D printers work?—A guide. Retrieved from https://3dprint.com/82272/what-3d-printing-works/

Lewis, P., & McVeigh, K. (2013, June 10). Edward Snowden: What we know about the source behind the NSA files leak. *The Guardian.* Retrieved from http://www.theguardian.com/world/2013/jun/11/edward-snowden-what-we-know-nsa

Lorenz, C. (1990). *The design dimension.* Oxford: Blackwell Publishing Ltd.

Lunden, I. (2015, July 2). 6.1B smartphone users globally by 2020, overtaking basic fixed phone subscriptions. Retrieved from http://techcrunch.com/2015/06/02/6-1b-smartphone-users-globally-by-2020-overtaking-basic-fixed-phone-subscriptions

Mahen, M. (2014, December 2). Over 151 Google products and services you probably don't know. Retrieved from http://www.minterest.org/google-products-services-you-probably-dont-know/

Maslow, J. (2015, February 15). China's big dilemma: Increasing labor costs. Retrieved from http://www.streetwisejournal.com/chinas-big-dilemma-increasing-labor-costs/

McCarron, J. (2007, 9 25). Marketing: Just do it! Turning slogans into sales. Retrieved from Pantagraph.com: http://www.pantagraph.com/business/local/marketing-just-do-it-turning-slogans-into-sales/article_fbb7041b-30f2-5e64-99bb-e7a249195aa1.html

McDermott, M. (2010, October 5). Is noise really why Sunchips should ditch bioplastic packaging? Retrieved from http://www.treehugger.com/corporate-responsibility/is-noise-really-why-sunchips-should-ditch-bioplastic-packaging.html

McDonald, G. (2016, April 8). Metal foam armor disintegrates bullets. Retrieved from http://news.discovery.com/tech/gear-and-gadgets/metal-foam-armor-disintegrates-bullets-160408.htm

McInnes, W. (2014, December 18). Listen up: Social listening proves critical for successfully launching new gadgets. Retrieved from http://venturebeat.com/2014/12/18/listen-up-social-listening-proves-critical-for-successfully-launching-new-gadgets/

McVey, T. (2011, June 30). ITAR—What government contractors need to know. Retrieved from http://www.williamsmullen.com/news/itar-%E280%93-what-government-contractors-need-know

Meulen, R. V. (2008, June 23). Gartner says more than 1 billion PCs in use worldwide and headed to 2 billion units by 2014. Retrieved from http://www.gartner.com/newsroom/id/703807

Morris, D. Z. (2015, August 27). Will tech manufacturing stay in China? *Fortune*, VOLUME, PAGES. http://fortune.com/2015/08/27/tech-manufacturing-relocation/

Morrissey, J. (2016, March 2). Sugru, a versatile glue from Ireland, gets help from web. The *New York Times*. Retrieved from http://www.nytimes.com/2016/03/03/business/smallbusiness/sugru-a-versatile-glue-from-ireland-gets-help-from-web.html

Murphy, S. (2011, April 13). $600 in groceries for $10? Yes. How? Extreme couponing. *The Christian Science Monitor*. Retrieved from http://www.csmonitor.com/Business/Latest-News-

Wires/2011/0413/600-in-groceries-for-10-Yes.-How-Extreme-couponing

Nechamkin, S. (2015, 9 22). How Shopkins became the biggest tiny toy on the planet. Retrieved from Racked.com: http://www.racked.com/2015/9/22/9365361/shopkins-tiny-toy-for-girls-consumerism-pink

Njord. (2016, January 11). Intel sues SeaIntel for trademark violation. Retrieved from http://www.lexology.com/library/detail.aspx?g=ffb48c13-da56-45b9-851c-7069a901b99e

Peterson, H. (2015, October 15). A war is breaking out between McDonald's, Burger King, and Wendy's—and that's great news for consumers. Retrieved from http://www.businessinsider.com/fast-food-price-wars-2015-10

Piombino, K. (2013, June). Listening facts you never knew. Retrieved from http://www.prdaily.com/Main/Articles/Listening_facts_you_never_knew_14645.aspx

Poeth, D. (2010, December 5). What employers need: Students need to prepare now for a tough job market. Retrieved from http://poeth.com/What_employers_need.pdf

Quirk, M. (2014, 7 19). 15 Product trademarks that have become victims of genericization. Retrieved from Consumerist.com: https://consumerist.com/2014/07/19/15-product-trademarks-that-have-become-victims-of-genericization/

Radjou, N., Prabhu, J., Ahuja, S., & Roberts, K. (2012). *Jugaad innovation: Think frugal, be flexible, generate breakthrough growth.* San Francisco, CA: Jossey-Bass.

Raju, J., & Zhang, Z. J. (2010). *Smart pricing: How Google, Priceline, and leading businesses use pricing innovation for profitability.* Indianapolis, IN: FT Press.

Rayport, J., Jaworski, B., & Breakaway Solutions Inc. (2003). *Introduction to e-commerce* (2nd ed.). Boston, MA: McGraw-Hill/Irwin.

Renault 1, (n.d.). Retrieved March 29, 2016 from Wikipedia: https://en.wikipedia.org/wiki/Renault_12

Robins, B. (2015, 11 25). 10 wow-worthy inventions of 2015. Retrieved from NY Daily News: http://www.nydailynews.com/news/world/10-wow-worthy-inventions-2015-article-1.2446673

Roy, V. (2015, June 30). 11 high margin products to sell online. Retrieved from http://www.insidermonkey.com/blog/11-high-margin-products-to-sell-online-357653/

Royer, Z. (2012, June 20). Nikola Tesla's wireless electric automobile explained. Retrieved from http://www.apparentlyapparel.com/news/nikola-teslas-wireless-electric-automobile-explained

Sherman, A. J. (2016, August 14). Protecting trademarks I: Definition. Retrieved from http://www.entrepreneurship.org/resource-center/protecting-trademarks-i-definition.aspx

Shiozawa, M. (2015, February 20). 16 things you didn't know about vending machines in Japan and around the world. Retrieved from http://www.coca-colacompany.com/stories/16-things-you-didnt-know-about-vending-machines-in-japan-and-around-the-world

Stansberry, G. (2010, February 25). 10 business leaders you should strive to emulate. Retrieved from http://www.businessinsider.com/10-examples-of-excellent-business-leadership-2010-2

Strauss, V. (2013, February 25). Why not to register copyright for unpublished work. Retrieved from http://accrispin.blogspot.com/2013/02/why-not-to-register-copyright-for.html

Sturm, F. (2015, October 18). How to recognize and protect your intellectual property. Retrieved from http://hsllp.com/protect.html

Tablet computer. (n.d.). Retrieved May 10, 2016 from Wikipedia: https://en.wikipedia.org/wiki/Tablet_computer

Toastmasters International. Who we are. Retrieved August 28, 2016 from http://www.toastmasters.org/About/Who-We-Are

Tool manufacturers. (n.d.). Retrieved March 8, 2016 from Wikipedia: https://en.wikipedia.org/wiki/List_of_tool_manufacturers

Trademarks. (n.d.). Retrieved April 16, 2016 from Wikipedia: https://en.wikipedia.org/wiki/List_of_generic_and_genericized_trademarks

Trott, P. (2008). *Innovation management and new product development.* Harlow, England: Pearson Education Limited.

Ulrich, K. T., & Eppinger, S. D. (2000). *Product design and development* (2nd ed.). Boston, MA: McGraw-Hill.

Urban, G. L., & Hauser J. H. (1990). *Design and marketing of new products.* Englewood Cliffs, NJ: Prentice-Hall.

US Patent and Trademark Office. Trademark class headings and explanatory notes. Last modified December 31, 2012. http://www.uspto.gov/trademark/trademark-updates-and-announcements/nice-agreement-tenth-edition-general-remarks-class

VentureNet. (2016). VentureNet evaluation form. Retrieved from http://www.iowaeconomicdevelopment.com/userdocs/documents/ieda/VentureNetCoEvalForm.pdf

Wilson, L. (2013, January 26). The difference between ABS and PLA for 3-D printing. Retrieved from http://www.protoparadigm.com/news-updates/the-difference-between-abs-and-pla-for-3d-printing/

Winch, G. (2013, June 18). 10 signs that you might have fear of failure. *Psychology Today*, VOLUME, PAGES. https://www.psychologytoday.com/blog/the-squeaky-wheel/201306/10-signs-you-might-have-fear-failure

Wright, H. (2015, January 28). Global power tool demand to hit US$33 billion. Retrieved from http://www.khl.com/magazines/international-rental-news/detail/item104647/Global-power-tool-demand-to-hit-US$33-billion

RECOMMENDED READING LIST

Design and Product Development

- Maslow's hierarchy of needs: http://psychclassics.yorku.ca/Maslow/motivation.htm
- Innovation advice from the UK government: Innovation.gov.uk
- Bad design examples: http://www.baddesigns.com/mirror.html
- Crazy ideas and inventions: Halfbakery.com
- Design Council: design-council.org.uk
- EU innovation: Europa.eu.int/comm/enterprise/innovation
- InnovationTools.com
- R&D Society: rdsoc.org
- Manufacturing: Stanford.edu
- Best stores for prototype building:
  - McMaster-Carr: http://www.mcmaster.com
  - Grainger: http://www.grainger.com
  - Digikey: http://www.digikey.com
  - Online Metals: http://www.onlinemetals.com
  - eBay: http://www.eBay.com
  - Amazon: http://www.Amazon.com
- Freelancer website: http://www.Upwork.com

Legal Protection of Ideas

- Patent basics: http://www.uspto.gov/patents-getting-started/ patent-basics/types-patent-applications/provisional-application-patent
- Drafting a patent application—a guide for inventors: http://www.canosoarus.com/16InventorTips/Patents.htm
- Responding to office action on your patent application: http://www.uspto.gov/trademarks-maintaining-trademark-registration/responding-office-actions
- Nondisclosure agreement template: https://www.score.org/resources/non-disclosure-agreement-template
- Basic nondisclosure agreement template: http://www.ndasfor-free.com/NDAS/GetBasic.html

Licensing the Product Idea

- Tips for licensing the idea: http://www.inc.com/stephen-key/how-to-license-an-idea-to-a-big-company.html
- "10 Reasons Why Companies Decide Not to License an Idea": http://www.inc.com/stephen-key/10-reasons-why-companies-decide-not-to-license-an-idea.html
- Top 150 global licensors: http://www.licensemag.com/license-global/top-150-global-licensors
- Inventor-friendly companies: http://idmagazine.wpengine.com/inventor-friendly-companies/
- Companies looking for product ideas: http://www.inventright.com/links/

Personal Development

- Rath, T., & Conchie, B. (2008). *Strengths based leadership: Great leaders, teams, and why people follow.* New York, NY: Gallup Press.

- Buckingham, M., & Coffman, C. (1999). *First, break all the rules: What the world's greatest managers do differently.* New York, NY: Simon & Schuster.
- Listening facts: http://d1025403.site.myhosting.com/files.listen.org/Facts.htm
- Dlugan, A. (2010, August 22). *17 easy ways to be a more persuasive speaker.* Retrieved from http://sixminutes.dlugan.com/logos-examples-speaking/
- Winch, G. (2013, June 18). *10 signs that you might have fear of failure.* Retrieved from https://www.psychologytoday.com/blog/the-squeaky-wheel/201306/10-signs-you-might-have-fear-failure

Crowd-Funding Websites

- Kickstarter.com is an all-or-nothing platform. (That means you don't get any money if you don't reach your goal.)
- Indiegogo.com and GoFundMe.com have flex funding. (That means you get everything you raise.)

Pricing

- "10 Examples of Great Pricing Strategies": http://upliftroi.com/blog/post/10-examples-of-great-pricing-strategies
- "Best Pricing Strategies": https://www.helpscout.net/blog/pricing-strategies/

Business and Marketing Plans

- Covello, J. A., & Hazelgren, B. J. (1994). *The complete book of business plans.* Naperville, IL: Sourcebooks.
- Kotler, P., & Armstrong, G. (2006). *Principles of marketing.* Upper Saddle River, NJ: Pearson Prentice Hall.

Online Retailers for Selling New Products

- **Amazon:** http://www.amazon.com
- **eBay:** http://www.eBay.com
- **Etsy** for artworks or creative works: http://www.Etsy.com
- **Pinterest** for creative works: http://www.Pinterest.com
- **Shopify** for opening online stores under monthly plans: http://www.Shopify.com
- **Bonanza** for the fashion industry: http://www.Bonanza.com

Advertisement and Sales

- Completely free advertising sources:
  - **Craigslist**, advertisement site: http://www.Craigslist.org
  - **Blogger**, blog website for advertising products: http://www.Blogger.com
  - **Facebook**, social networking website: http://www.FaceBook.com
  - **Twitter**, social networking site: http://www.Twitter.com
  - **Reference.com**, reference site: http://www.Reference.com
  - **MySpace**, social networking site: http://www.MySpace.com
  - **YouTube**, video-sharing site: http://www.Youtube.com
  - **DMOZ**, most search engines get information here: http://www.DMOZ.org
- Sales training and resources:
  - Websites
    - **LinkedIn** (www.LinkedIn.com), **Quora** (http://www.quora.com/Sales), and **Focus** (http://www.focus.com/topic/sales/): all three of these social networks contain great discussion areas for sales experts to ask and answer sales-related questions.
    - **Inbound.org:** a social news site where sales and marketing professionals curate and share the best content from the web: http://inbound.org/

- **Alltop Sales:** an aggregator that pulls in headlines from all the popular sales blogs and displays them on one page: http://sales.alltop.com/
- **Sales Gravy:** a social network for sales professionals to share opportunities and resources and connect with other like-minded professionals: http://www.salesgravy.com
- **Famous people** who were rejected for their product ideas but didn't give up: http://www.uky.edu/~eushe2/Pajares/OnFailingG.html
- Books
  - Cialdini, R. (1993). *Influence: the psychology of persuasion.* New York, NY: Collins.
  - Rackham, N. (1988). *Spin selling.* New York, NY: McGraw-Hill.
  - Carnegie, D. (2009). *How to win friends and influence people.* New York, NY: Simon & Schuster.
  - Holmes, C. (2008). *The ultimate sales machine: Turbocharge your business with relentless focus on 12 key strategies.* New York, NY: Portfolio.
  - Hopkins, T. (1982). *How to master the art of selling.* Scottsdale, AZ: Champion Press.

INDEX